DOMINIC CHINEA

MACHINES

A VISUAL HISTORY

DOMINIC CHINEA

MACHINES

A VISUAL HISTORY

Illustrated by LEE JOHN PHILLIPS

100 MACHINES AND THE REMARKABLE STORIES BEHIND EACH INVENTION

Editor Millie Acers
Designer James McKeag
Production Editor Marc Staples
Senior Production Controller Louise Minihane
Senior Acquisitions Editor Pete Jorgensen
Managing Art Editor Jo Connor
Managing Director Mark Searle

Text Dominic Chinea and Nathan Joyce
Illustrations Lee John Phillips
Project and Jacket Designer Eoghan O'Brien

DK would like to thank Caroline Curtis
for proofreading and indexing.

First published in Great Britain in 2024 by
Dorling Kindersley Limited
DK, One Embassy Gardens, 8 Viaduct Gardens, London SW11 7BW

The authorised representative in the EEA is
Dorling Kindersley Verlag GmbH. Arnulfstr. 124,
80636 Munich, Germany

Page design copyright © 2024 Dorling Kindersley Limited
A Penguin Random House Company
10 9 8 7 6 5 4 3 2 1
001–338111–Nov/2024

Text copyright © 2024 Dominic Chinea
Artwork copyright © Lee John Phillips, 2024

All rights reserved.
No part of this publication may be reproduced, stored in or introduced
into a retrieval system, or transmitted, in any form, or by any means
(electronic, mechanical, photocopying, recording, or otherwise),
without the prior written permission of the copyright owner.

A CIP catalogue record for this book is available from the British Library.
ISBN: 978-0-2416-4554-3

Printed and bound in China
www.dk.com

CONTENTS

Introduction	6
1 **ANCIENT MACHINES**	10
2 **FOOD AND DRINK**	44
3 **HERITAGE CRAFT**	74
4 **THE VICTORIAN AGE**	120
5 **FAIRGROUND AND NOVELTIES**	166
6 **HOME AND RECREATION**	200
Index	234
About the Author	239
Acknowledgements and Picture Credits	240

INTRODUCTION

Many machines make life easier, solve problems or save time. Some of them have changed people's lives or saved their lives. A few of them have even changed the course of history. And some of them are just a bit of fun!

Whatever the reason for their creation, each of the 100 machines I've chosen for this book are unique accomplishments. Some are particularly associated with a country's cultural heritage. Others, like the sewing machine or the Ferris Wheel, feature stories fit for a film script. There are tales of triumph against all odds, marketing magicians, unique partnerships and tragic stories of unsung geniuses who just couldn't catch a break. I've learned some remarkable things writing this book. I'll never look at a treadmill the same way again, I can tell you that!

Just sitting in here looking around my workshop, I'm surrounded by as many machines as I am tools. Some of them have become an indispensable part of a craft and a big part of my life. In exactly the same way that I'm passionate about using hand tools over power tools, the machines I love the most are manually operated. Nothing beats cranking a lever, which turns the gears and brings that machine alive. That's why most of the machines in this book are powered by hand or foot, except for a few steam-powered masterpieces, hydraulic and pneumatic game-changers and the very occasional internal combustion engine that I couldn't resist.

Each of these machines represents a remarkable achievement. It means that someone has had a vision and gone on a journey of discovery, finding the courage and self-belief to see it come to life. No doubt, they would have been questioned and criticised along the way. They would have to contend with the frustration of trying and failing, trying and failing, over and over again. Winston Churchill said: "Success is the ability to go from failure to failure without losing your enthusiasm" and that really struck a chord with me. It's so much easier to give up than to keep going, so you've got to have shedloads of resilience. You've also got to be pretty stubborn, single-minded and not easily offended!

I'm drawn to the machines that are more than just functional designs. That's why I love the Ranalah wheeling machine, the Wilcox & Gibbs sewing machine or the Harrington's dental drill. They've each been made by craftspeople who have gone a step further to make something beautiful, elegant, sleek or delicate. Sometimes they've added artistic flourishes that transport you to a time when machines were being exhibited for the first time, at a groundbreaking event like the Great Exhibition of 1851. But sometimes these remarkable details are hidden to everyone but the people who repair these incredible machines. It's when you open them up that you discover a hidden language handed down between craftsperson to craftsperson generations apart.

Tinkering with mechanical things began with a skateboard for me, which basically involved taking off the nut that holds the axles on. Then it progressed to fixing my pushbike with its bigger wheels, spokes, tyres and brakes. Before I knew it I was repairing my mountain bike and dealing with chain tension and gears. Next thing you know, you're rebuilding the suspension and modifying this part and that. In turn, that's a gateway to buying a basic bicycle frame and curating the components yourself. Then you move on to a motorbike with an engine and the sheer number of things you can tinker around with is mind-blowing. You gain more mechanical knowledge at each stage and the whole thing becomes addictive. So eventually you end up with a workshop full of extraordinary machines that need fixing and a list that keeps getting longer and longer. And I absolutely love it.

DOMINIC CHINEA

ANCIENT MACHINES

MACHINE NUMBER	MACHINE NAME	PAGE NUMBER
001–006	THE SIX SIMPLE MACHINES	014—017
007	GEAR TRAIN	018
008	ARCHIMEDES' SCREW	019
009	WATER WHEEL	020
010	SPINNING WHEEL	022
011	WINDMILL	023
012	POTTER'S WHEEL	024
013	CRANE	026
014	PLOUGH	027
015	PRINTING PRESS	028
016	WHEELBARROW	030
017	TREBUCHET	032
018	CATAPULT	034
019	BELLOWS	035
020	PENDULUM CLOCK	036
021	MECHANICAL CALCULATOR	038
022	SURVEYOR'S WHEEL	040
023	SUBMARINE	042

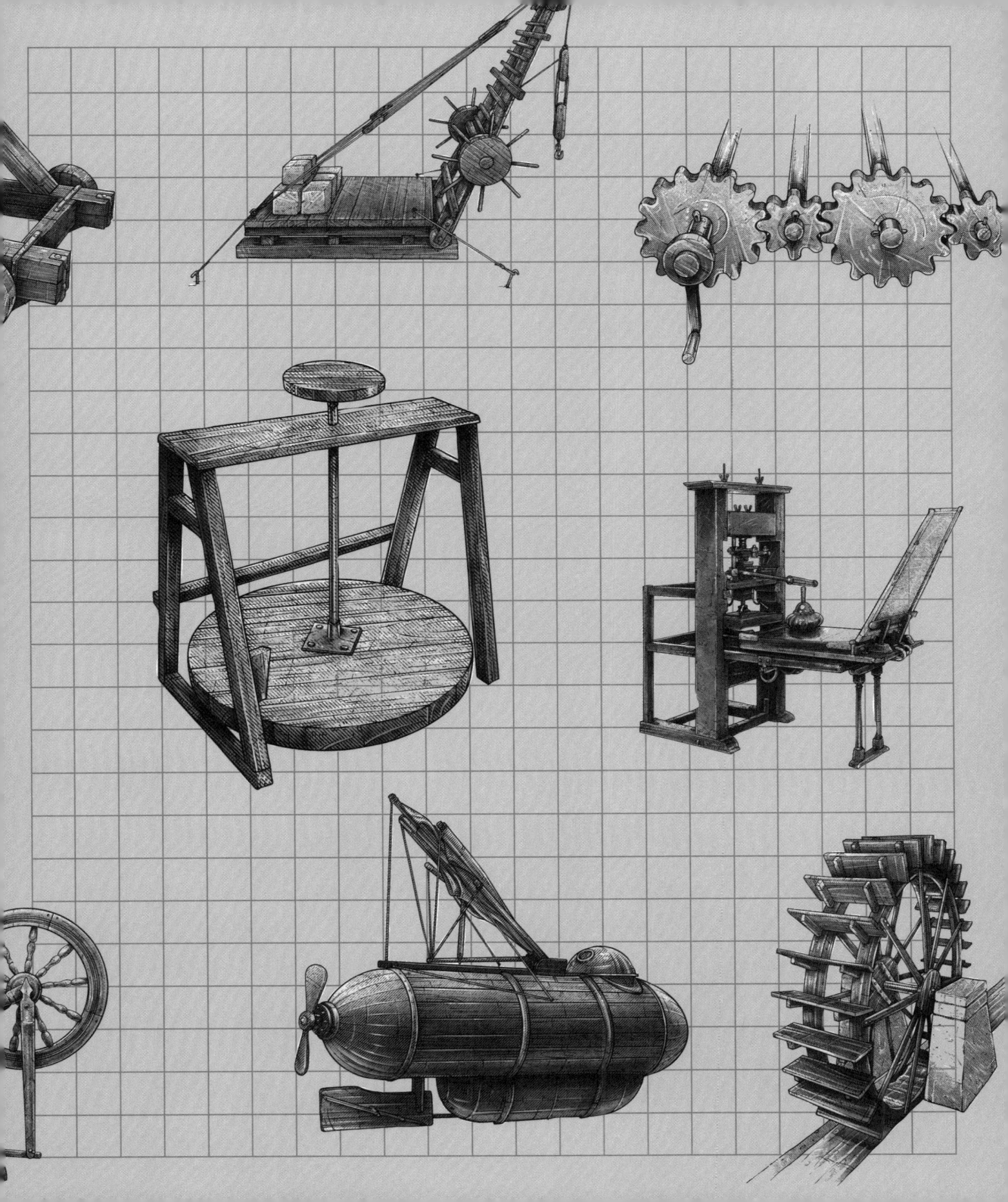

THE SIX SIMPLE MACHINES

There are six simple mechanical machines: the inclined plane, the wedge, the screw, the lever, the wheel and axle, and the pulley. They're all ancient machines, dating back thousands of years, and some of them don't look like what you might imagine when you hear the word "machine". So how do they work? Each of them reduces the amount of effort we need to use to help us perform a particular task. And they do this by increasing the amount, or changing the direction, of a force. Some more complicated machines, like a bicycle, are actually a combination of several simple machines.

INCLINED PLANE

This is literally just a sloped surface which supports an object's weight as it moves upwards along the slope. This means it requires less force to push the object compared with lifting it straight up. The drawback is that you need to apply the force over a longer distance. But that's a small price to pay when you're trying to move a piano, for example.

WEDGE

This is basically a moveable inclined plane with a thick end and a pointy end. If you've ever used an axe to cut wood for a fire, you're using a wedge. So if you apply force to the thick end of the wedge, say with a hammer, the pointy end and the sloping sides of the wedge will break or split apart an object. The force you've put in is multiplied by the wedge.

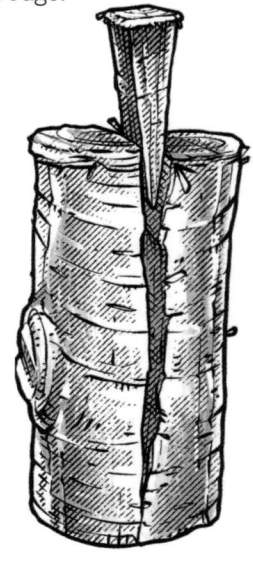

LEVER

A lever is a beam or bar that rotates around a fixed point called a fulcrum. The best example of a lever (to me, anyway) is when you are using the claw end of a hammer to pull a nail out of wood. You push down on the handle of the hammer, and the head of the hammer, which makes contact with the wood beneath, is the fulcrum. The force you've applied is transferred and increased at the claw end, which easily pulls out the nail. So a lever allows you to use a small force to lift something heavy. And speaking as someone who very regularly uses a hammer to pull out nails, I'm very grateful that I don't have to do that with my bare hands.

SCREW

A screw is really a type of inclined plane that wraps around a central cylinder. It allows you to convert rotary motion into forwards or backwards motion. They're used both to fasten objects and to move material. The legend goes that Archytas of Tarentum invented the screw around the 5th century BCE, and Archimedes' water screw dates back to the 3rd century BCE. And it seems pretty clear that the Ancient Egyptians were using similar machines long before this to irrigate crops. When you and I think of a screw, we think of a metal object with a thread – basically a ridge that sits on the outside of the cylinder. Metal screws were used from the 15th century CE, but I want to fast-forward to 1841 and Sir Joseph Whitworth, who suggested that all British screw manufacturers use a 55° thread-angle, a consistent thread-depth and consistent radius. It was the first national screw thread-standard in the world and it solved a lot of problems, as before then every imaginable screw, size-wise, length-wise and thread-wise, had been produced. The US followed suit in 1864, with William Sellers proposing a 60° thread-angle, and that became the US standard. And then in 1948, the UK and Canada also adopted the Sellers standard, so they were finally all on the same page.

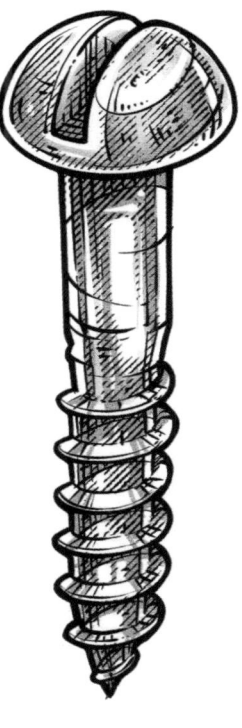

THE PULLEY CAN BE TRACED BACK TO ANCIENT EGYPT AND IS CLOSELY RELATED TO THE CRANE.

WHEEL AND AXLE

MACHINE NUMBER 005

A wheel with an axle – a rod that passes through the centre of the wheel – is a simple machine that alters force. It was probably used, at first, for lifting or lowering weights, like buckets into water wells. The wheel and axle move together and can be used in different ways, by applying force to either the axle or the wheel. With a doorknob and latch, the "wheel" is the doorknob and the "axle" is the thin shaft running through the centre. It requires less force to turn the knob, in order to open the latch, than it would do to turn the axle. It makes the job easier. Likewise, if you think about a Ferris wheel, turning the axle rotates the wheel using less force. It's much easier to turn the axle than it is to turn the wheel. Most of my life involves fixing objects with wheels, from go-karts to bicycles, soapbox racers to campervans, and even the machines and devices I use to fix those objects have wheels, like the wheeling machine (see page 104).

PULLEY

MACHINE NUMBER 006

A pulley is a wheel with a grooved rim around the outside which carries a cord, rope, chain or belt. A pulley is used either by itself or with multiple pulleys to transfer power and/or motion, typically to lift heavy loads.

The pulley can be traced back to Ancient Egypt and is closely related to the crane (see page 26), which would have originated from pulley systems. It's a truly ancient machine, going back to at least 1500 BCE in Mesopotamia (modern-day Iraq).

One pulley on its own doesn't offer what's called mechanical advantage, because you still have to use the same amount of force to move the load. What it does do, is change the direction of the force you need to apply (with a pulley, you're pulling a rope downwards, which is easier than lifting the load up with your hands). Pulleys become much more helpful when they're used together. If you've got two connected pulleys, the force you need to move the load is halved. If you've got three connected pulleys, three sections of rope take the strain, so the force you need to pull the rope is divided by three. A system of two or more pulleys is called a block and tackle, and it's something that Archimedes is thought to have pioneered. According to the Ancient Greek philosopher Plutarch, Archimedes boasted to King Hiero that any weight in the world could be moved. The King asked Archimedes to prove it, and Archimedes chose one of the king's ships to be the subject of the experiment. Here's how it is said to have gone: "loading her with many passengers and a full freight, sitting himself the while far off, with no great endeavour, but only holding the head of the pulley in his hand and drawing the cords by degrees, he drew the ship as smooth and evenly as if she had been in the sea." Now we don't know for sure that this happened, but it's a great story.

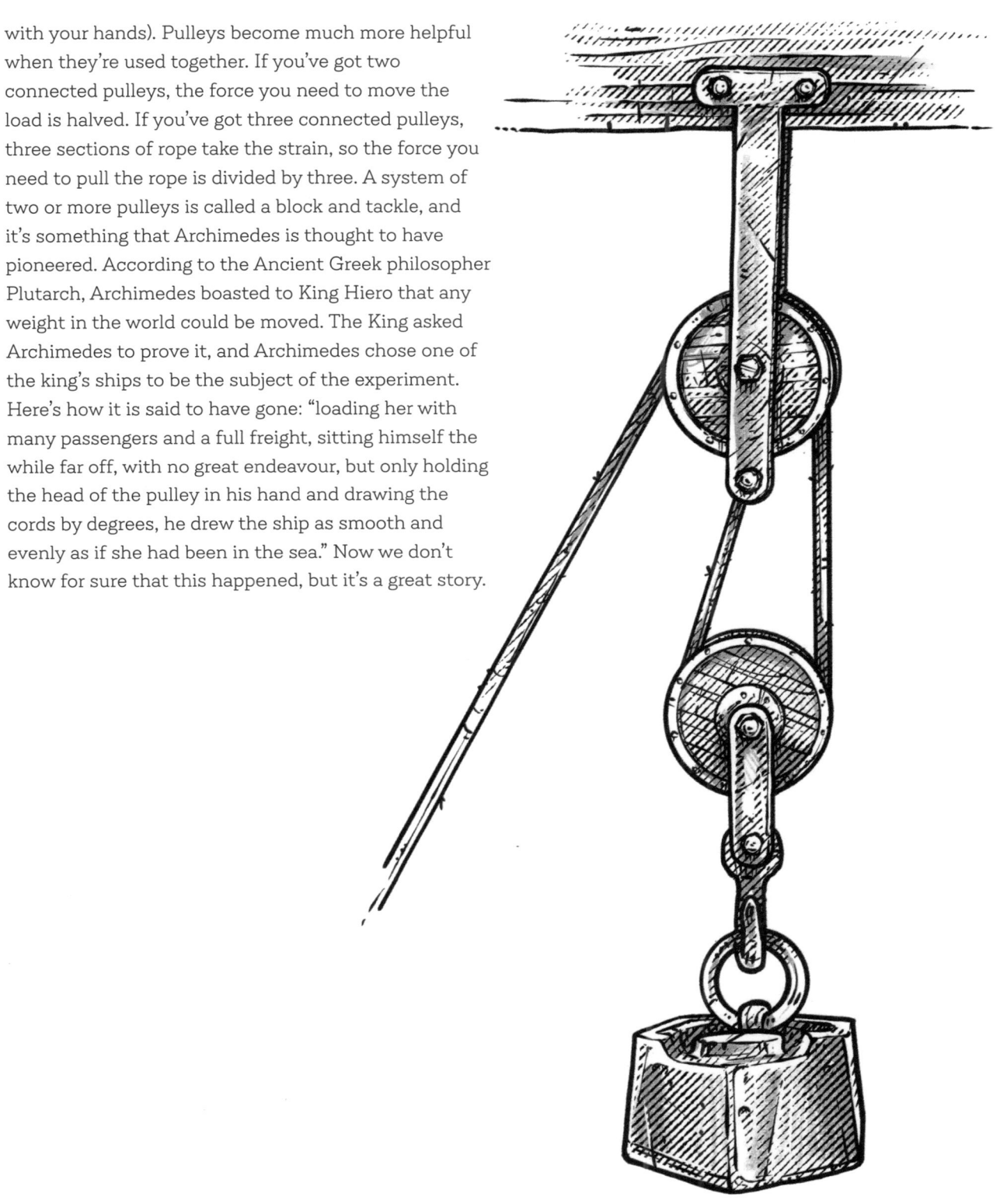

GEAR TRAIN

MACHINE 007

The gear train is a system of two or more gears (circular machine parts with teeth around the outside) that interlock, and smoothly transfer motion from one place to another.

They're used mainly to speed up or slow down the speed of rotation, to reverse the rotation, or move the motion to a different axis. Smaller gears containing fewer teeth turn faster than larger gears with more teeth, and the difference in speed is known as the gear ratio. If it's 3:1, each time the larger gear goes around once, the smaller one will go around three times.

You'll find gear trains in almost everything with spinning parts, especially those where engines, or motors, produce rotational motion. One of the most famous mechanical devices with a gear train was the Antikythera mechanism, and it's something straight out of an *Indiana Jones* film. Found in a shipwreck in the Mediterranean in 1901, it's an Ancient Greek astronomical calculator used to chart the movement of the sun, moon and known planets, and is believed to date back to 100 BCE. Technically, it's an analogue computer. Although found in fragments, it did feature at least 35 bronze gear wheels and 19 shafts and axles, and was probably operated via a small hand crank, which rotated all sorts of dials and scales featuring Greek Zodiac symbols. It's a truly incredible object, and the most complicated machine to survive from the Ancient World.

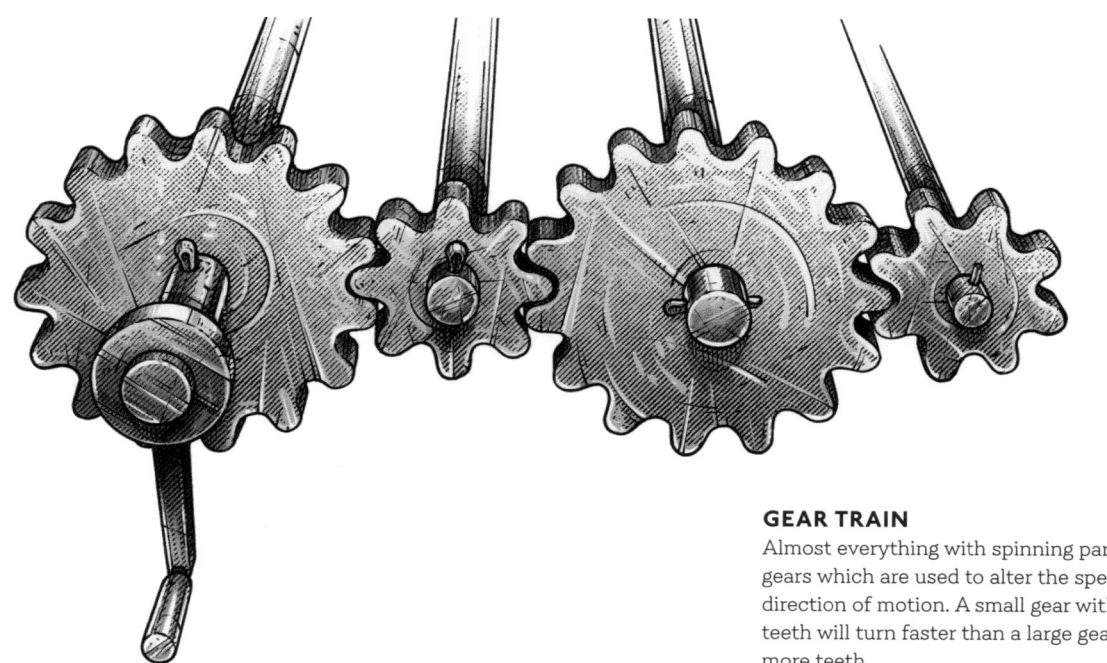

GEAR TRAIN
Almost everything with spinning parts uses gears which are used to alter the speed or direction of motion. A small gear with fewer teeth will turn faster than a large gear with more teeth.

ARCHIMEDES' SCREW

The Archimedes' screw, also known as the water screw, is a machine used to raise water. It consists of a huge spiral fixed inside a hollow tube, mounted at roughly a 45° angle, with the bottom of the tube submerged in water, and the top end, typically featuring a handle (in early versions of it anyway), which you rotate to scoop the water upwards.

It's a simple but utterly ingenious invention and is still commonly used today in pumps used at water-treatment plants and in hydroelectric power stations. It takes its famous name from the legendary Greek inventor and mathematician Archimedes, but it was most likely already being used in Ancient Egypt long before he was around to raise water from the River Nile to irrigate crops. Still, everyone likes a good story, and the supposed origin is that Archimedes was commissioned by King Hiero II of Syracuse to design the most magnificent ship in the world. Called *Syracusia*, the resulting ship, with gardens, temple and gymnasium, was also said to include this device for removing leaking water from the ship's hull.

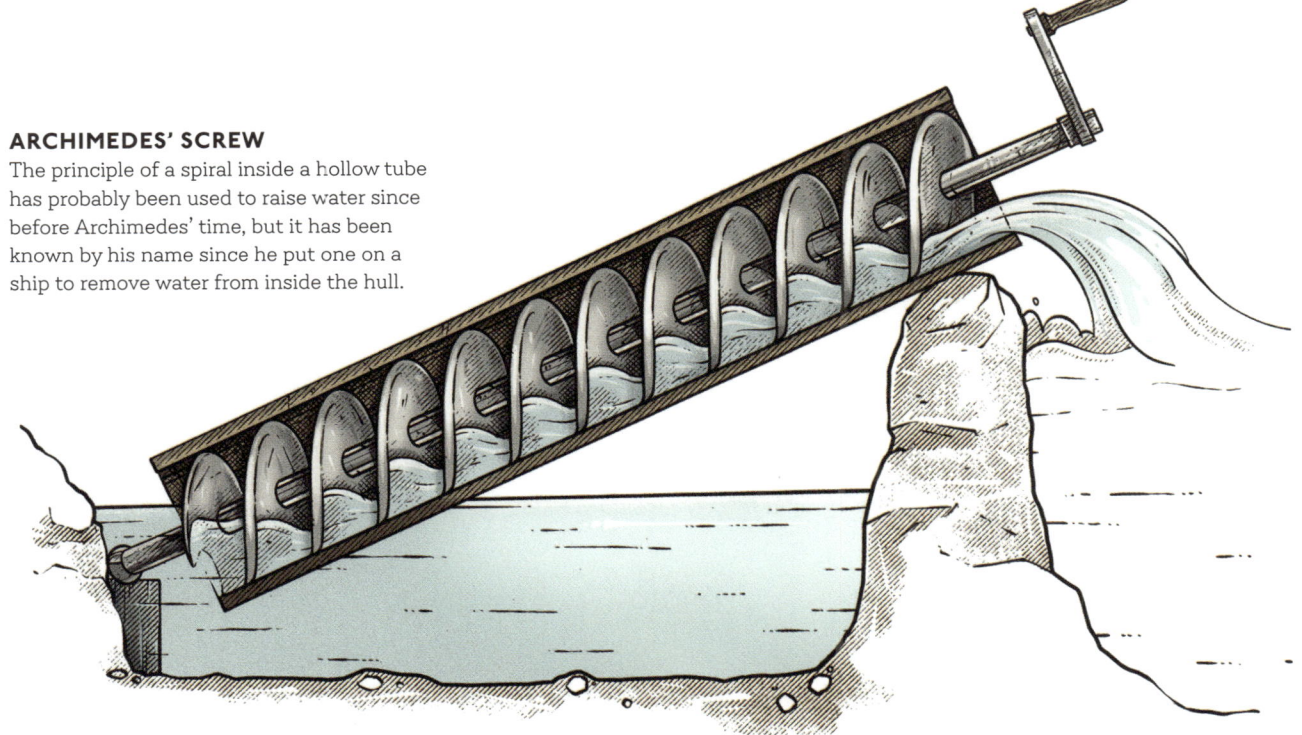

ARCHIMEDES' SCREW
The principle of a spiral inside a hollow tube has probably been used to raise water since before Archimedes' time, but it has been known by his name since he put one on a ship to remove water from inside the hull.

WATERWHEEL

MACHINE NUMBER 009

A waterwheel consists of a wheel mounted with blades, paddles or buckets, and uses the power of running water pushing against these to make it turn.

It's one of the earliest forms of mechanical power, and would have been initially used to lift water, but connecting the shaft of the wheel to machinery opened up a whole new world of mechanised possibility. This included grinding grain into flour (in what became known as a gristmill), crushing fibre to turn into cloth, and driving equipment like saws (sawmills). It's thought there were more than 20,000 waterwheels in the UK at the beginning of the 19th century, which had been adapted for many different uses, including powering trip-hammers in forges, crushing timber and grinding flint.

There are two main types of waterwheel – a horizontal wheel containing a vertical axle (sometimes known as a Norse mill or Greek mill); and a vertical wheel containing a horizontal axle. With a water-powered mill, you'd often find a dam, or weir, across a section of water to create a millpond. From this, engineers would construct a channel, called a millrace, to direct water into the waterwheel. A millpond is almost always calm and flat, hence the expression "calm as a millpond" (or just "like a millpond"), which was famously used by Captain Edward J. Smith of the *Titanic* on the fateful night the ship was lost.

Waterwheels were eventually superseded by the invention of the turbine in the 19th century, but they were still widely used until the end of the 19th century, and many still exist in developing countries. Some of the remaining ones in Europe and North America have been converted into mini hydroelectric plants, so as to preserve the beauty of the original machine and supply renewable energy.

There's a waterwheel and attached mill-building at the Weald & Downland Living Museum, where *The Repair Shop* is filmed, and I walk past it every day. The waterwheel is made of cast iron, is 3.66m (12ft) in diameter and uses around 900 litres (200 gallons) of water to complete one revolution. As with much of the remarkable machinery at the museum, it was constructed elsewhere centuries ago and was rebuilt on site. A team here had to excavate two ponds, an upper and lower one, and there's a height difference of approximately 3.66m (12ft) between the two. After the water runs from the upper pond, down the millrace and against the wheel, it runs to the lower pond and is circulated back to the upper pond by a powerful pump. The wheel itself drives a horizontal wooden watershaft, and this shaft is connected to a system of gears that drive grindstones. The best thing about the waterwheel and the mill is that it's working and is used to grind flour, which you can buy at the museum shop. It is such a beautiful structure, and the engineering inside the mill building is just mind-blowing. Showing people exactly how something works, watching everything turning perfectly, and seeing how it all connects to the grindstones that the miller uses to turn cereal grains into flour, is really inspiring. It brings heritage craft to life, witnessing such a remarkable process (and the end product being made), and knowing that it's been going on for thousands of years.

My good friend Nick, who runs the forge at the museum, volunteers one day a week at the mill. Nick knows instantly if something doesn't sound right, where the problem lies and what to do to fix it.

The mill has almost become an extension of him. I went to see him in spring 2023, when he was busy fixing the axle – and it's absolutely huge, like a whole tree trunk in size. The creaking sound it makes when it turns sounds like you're aboard some kind of galleon on the high seas, and you can feel the force of the whole mechanism running right through you when you're standing next to it. It's a real art having the knowledge to fix and maintain such an incredible machine, and he's passing on his knowledge to other volunteers he has inspired. And that's what I love – seeing age-old skills being transferred from an expert in their field to other folks that they've inspired.

Walking past that incredible waterwheel every day makes me feel lucky to work on a television show where everything around you isn't just a façade or something built temporarily for the purposes of filming. Everything here is real, works, and is maintained by people who want to showcase and preserve these incredible tools and machines.

WATERWHEEL
There were probably more than 20,000 waterwheels in the UK at the beginning of the 19th century. They were used for everything from grinding grain into flour to powering trip-hammers in forges.

SPINNING WHEEL

Before the spinning wheel was invented, people produced long continuous strands of thread (yarn) by manually drawing out and twisting a mass of animal fur, or plant fibres, held on a stick known as a distaff. They'd guide the yarn onto a spindle – basically a stick with a weighted disc at the end, and maybe a hook or groove to help guide the yarn. It would have taken absolutely bloody ages.

The spinning wheel wasn't an updated version of the spindle – it helped with the twisting process before the yarn was guided onto the spindle. Although its origins are much debated, the spinning wheel is thought to have originated in India, somewhere between 500 and 1000 CE, and reached the Middle East and then Europe by the 13th century. And it was in Europe three centuries later that a big development arrived in the form of the Saxon Wheel – this featured a fixed vertical rod instead of a separate distaff, a bobbin (a cylinder on which the thread wound continuously) and a foot treadle that turned the wheel, leaving both of the spinner's hands free. It sped things up a lot more because you didn't need to keep stopping to wind up the yarn.

Everything changed with the development of the spinning frame during the Industrial Revolution, which mechanised the process. The spinning jenny, invented around 1765 by English spinner and weaver James Hargreaves, was a game-changer, involving a metal frame and eight wooden spindles, which all revolved when the spinner turned a large wheel, spinning the thread. Lancashire entrepreneur Richard Arkwright patented his water frame in 1769, which used water power to drive it, and not only created stronger, harder yarn than the spinning jenny; it also could spin 96 threads simultaneously. Arkwright and his business associates went about building a huge factory next to the River Derwent. He built mills across Derbyshire and Lancashire and despite having both of his patents revoked in a trial in 1785, he became a very successful businessman, and hundreds of factory owners followed his example. He was known as the "father of the factory system".

WINDMILL

MACHINE 011 NUMBER

The first reference to a windmill is in the works of Persian geographer Estakhri in the 9th century CE, and it's possible that early versions existed from the 7th century CE.

They didn't look like how you're probably imagining, though – instead of being fixed on a horizontal axis like a modern wind turbine, they were vertical axle windmills, mounted in a fixed building. They worked a bit like fixed spinning tops. Incredibly, there are still operational examples of this type of windmill in northeast Iran which are at least 1,000 years old. The Nashtifan windmills are a series of 20m/65ft-high structures made of wood, clay and straw which look a bit like massive, full-height turnstiles at sports stadiums. The vertical mast runs down to a huge rotating grindstone in a room beneath, which ground cereals into flour. Incredibly, they're still operational and are (as at 2023) looked after by an ancient caretaker called Ali Muhammed Etebari.

The more familiar horizontal-axis windmill with sails radiating outwards was inspired by the Roman watermill and appeared in Europe in the 12th century. The earliest type is known as a post mill, so called because the entire body of the mill (housing the grinding stones and gearing) rotates around a central vertical post so it can move with the wind. It typically has four sails, which were wooden frames covered in sailcloth. The sails are mounted on an axle, called a windshaft. Stopping a windmill was quite a basic operation at first, involving a miller standing at ground level with a rope wound around the windshaft, but that all changed when brake wheels were developed and fitted onto the windshaft behind the front bearings. There's a well-known post mill in Outwood, Surrey, not far from London, which was built in 1665. The original owner, Thomas Budgen, is said to have seen the Great Fire of London (which happened in 1666) from its partially completed roof. It is thought to have last been used in 1996 to grind corn, and it came up for sale in 2018 for £800,000.

WINDMILL
The first recorded windmills had sails fixed on a vertical, rather than horizontal, axis. The more typicals one that we still see today were inspired by the Romans and arrived in Europe in the 12th century.

POTTER'S WHEEL

MACHINE NUMBER 012

It's hard to even imagine how pottery was shaped before the invention of the potter's wheel, but clay would have been rolled, pinched and smoothed in what must have been a really time-consuming, and knackering, process.

We know that the Ancient Sumerians in Mesopotamia were using stone potter's wheels as early as 3250 BCE. They were also being used in Egypt around this time, and in Ancient Egyptian mythology, Khnum, the God of the source of the River Nile, created human children on a potter's wheel from clay.

Potter's wheels would have been simple at first – a circular stone platform connected to a much narrower rounded base that could be manually turned with one hand while the potter worked with the other. With the appearance of the flywheel and turntable-shaft during the Iron Age, the process would have become much easier, freeing up both the potter's hands and making the whole thing a smoother, more predictable process.

Potters tend to use their left hand as their stabilising hand for roughly shaping the clay, while their dominant hand is used for the more intricate finer detailing.

The potter's wheel itself has changed very little – electric motors tend to move the wheel now, but what certainly hasn't changed is the centrifugal force (harnessed by the potter, which is what makes the clay expand outwards while it's being turned).

A potter's wheel arrived on my workbench for *The Repair Shop* recently, so I'll get to restore one to its former glory. It belonged to a well-known ceramicist who passed away and has ended up with his children, who fondly remember the smell of the clay and the sound of their dad making pot after pot. The sound of the motor became like the heartbeat of the entire house and they used to cook jacket potatoes in the kiln while it was cooling down. Hearing all that reminded me how much I've always wanted to use a potter's wheel, but somehow it has never happened. I've been to lots of pottery fairs, watched people in front of me making pots, and got really into *The Great Pottery Throw Down*. We've had amazing potters on *Make it at Market* and Florian (the ceramicist on the show) even brought down two potter's wheels, but I got to where they were filming him too late, after they'd packed everything away. I'll get there one day.

One thing I didn't know about potter's wheels is that many are reversible – they can spin either clockwise or anti-clockwise thanks to an electric switch. Most people learn with the wheel spinning anticlockwise. I can't pretend to be an expert on this but I have heard that right-handed people tend to find it more comfortable with the wheel spinning anticlockwise, while lefties tend to prefer the wheel to spin the other way. But it seems to be very much a

THE POTTER'S WHEEL ITSELF HAS CHANGED VERY LITTLE.

personal choice for the potter, with some potters choosing to rotate the wheel one way for throwing (the initial shaping of the clay) and another for trimming (removing excess clay and refining the item). In Japan, China and South Korea, potter's wheels usually turn clockwise but I haven't found a definitive answer as to why they do it differently there. I have read a claim about it having something to do with valuing the inside of the pot in the East more than the outside, as we seem to in the West, which is a fascinating point, but I can't verify it. Trying to find the answer led me down a rabbit hole about *Kintsugi*, the Japanese art of repairing broken pottery with powdered gold, silver or platinum. Not only does this make cracks beautiful; it also draws attention to them, embracing them as part of the history of the object. There's beauty in imperfection.

POTTER'S WHEEL
The arrival of the turntable shaft and flywheel in the Iron Age would have freed up the potter's hands and made the process of pot-making much smoother.

CRANE

```
One of the first known cranes was the shaduf, a hand-
operated device that lifts water from a water source,
usually into some kind of receptacle on land. It was
used mainly as a method of irrigating crops, and dates
back to around 3000 BCE in Mesopotamia.
```

MACHINE NUMBER 013

A *shaduf* is basically a really long pole on a pivot that's set on top of a wooden ladder so the actual pivot is several feet in the air. The pole extends more to one side than the other, though – at the end of the longer side is a rope attached to a bucket, while the shorter side has a weight. You pull down on the rope and the counterweight lifts the bucket back up. It's simple, clever and such a good invention that it's still in use in parts of the world today.

As for construction cranes, we know that early forms of these were around in Ancient Greece in the 6th century BCE thanks to the distinctive grooves along the bottom and sides of large stones in Greek temples in the ancient cities of Corinth and Isthmia. They must have been used to secure ropes that would have been attached to a lifting device, although we don't know exactly what it would have looked like. The compound pulley system followed. The Romans built on these technological leaps, developing the treadwheel crane, which allowed much heavier weights to be lifted. And when I say treadwheel, I mean a huge wheel that was forced into motion by a person's feet and hands and connected to a rope and pulley system.

The first mechanically powered crane was the steam crane, i.e. a crane powered by a steam engine. Some of these were mounted on simple railway cars and designed to travel to the scene of a breakdown. They became commonly known as "accident cranes", and were first made in 1875 by Appleby Bros – based in Southwark, London (with another works up in Leicester), the company became famous for its cranes and construction machinery. It had exhibited its early steam cranes at the International Exposition of 1867 in Paris, and sold its first accident cranes to Midland Railway. Almost every other railway company in the country was ordering accident cranes by the 1890s.

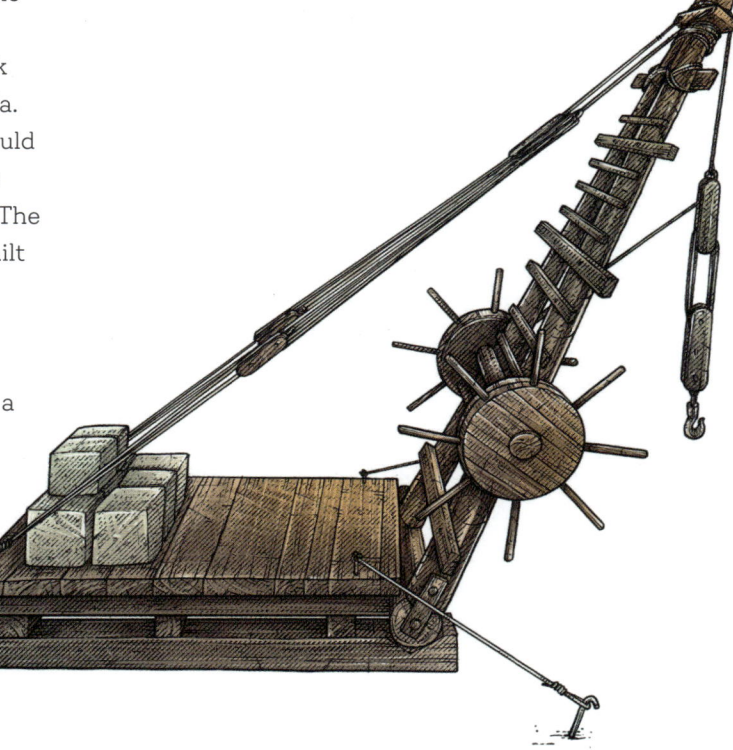

PLOUGH

MACHINE NUMBER 014

The plough is the tool that kick-started civilisation. Around 12,000 years ago, humans were nomadic hunter-foragers, but the world started to warm up slightly, and as things got hotter and drier, animals abandoned higher places in favour of river valleys.

Humans followed, and they settled there and began to farm. And to do that, the first thing you need is something to break up the top soil to bring nutrients up and allow moisture to sink down. At first, humans probably just used sharpened sticks, eventually adding handles. Then came along the animal-driven scratch plough. This had a wooden cutting blade, known as a share, which gouged a hole into the ground, and a handle (stilt) at the other end to help steer. By the time of the Romans, ploughshares would have been made of iron. Around 2,000 years ago, the mouldboard plough appeared in China, but it would be centuries until it arrived in Europe. The mouldboard was a curved plate attached to the ploughshare, which lifted and turned the soil after the share had gouged a hole. The coulter was also eventually added, a vertically mounted blade that cut into the soil ahead of the ploughshare and made cultivating heavier soils much simpler. Wheeled ploughs were developed, which helped with manoeuvrability and control.

The basic design of the plough changed remarkably little until the 18th century. In 1730, Jospeh Foljambe invented the Rotherham swing plough, with a mouldboard made of iron. The design was a lot lighter and easier to use than other ploughs. Across the Atlantic, blacksmith John Deere was getting fed up with having to repair his cast-iron and wood plough that was struggling in the heavy and sticky soils of the eastern US. So in 1836, he invented a much stronger cast-steel plough that tilled the soil without getting clogged up. By 1846, he'd established a business that was turning out 1,000 ploughs a year. The advent of steam power was harnessed around 1850 with the balance plough, which could plough multiple furrows. But these wouldn't last that long, as the turn of the century brought with it a mechanised game-changer: the wheeled tractor. And then Harry Ferguson's 3-point linkage system in 1920 revolutionised the way agricultural implements were attached. His name lives on in the famous brand Massey Ferguson.

PRINTING PRESS

Johannes Gutenberg was the German craftsman who created a printing press, around 1436, which was a technological game-changer.

He wasn't the first to use moveable metal blocks, though – Korean bookmakers had been doing that since at least the 13th century. The story goes that a chap called Ch'oe Yun-ŭi was commissioned to print a long Buddhist text, which would have involved making thousands of individual wooden blocks. Choe developed a time-saving solution by coming up with moveable metal type. He did this by carving a character into a piece of wood, which he then pressed into a little sand trench, and then filled that with molten metal. But the printing part of the process all took place by hand.

What Gutenberg did was to come up with a type-mould for each letter. You pour in the molten metal (he developed an alloy that both melted and cooled down quickly) and you'd have the shape of your letter. He also developed thick, oil-based inks that would both adhere to the metal type and transfer to vellum (parchment made from calf skin) or paper. His printing press was inspired by ancient olive and wine presses and featured a handle that would turn a strong wooden screw, which would force the upper platen (a flat plate) downwards to meet the paper or parchment laid over the moveable type, which was mounted on another plate (the lower platen). The first book he printed, in 1455, was the *42 Line Bible*. Book printing had been properly mechanised, and it was one of the most significant technological developments in history. His press would eventually print 250 pages an hour, which is pretty incredible, especially when you consider that the alternative was copying an existing manuscript using a pen that you'd have to keep dipping in ink.

One hundred and eighty copies of Gutenberg's Bible are thought to have been printed – 135 on paper and 45 on vellum. Only 48 copies remain, although most of them are incomplete. According to the British Library, only twelve paper copies and four vellum copies are complete. Amazingly, seven of them are in the UK – the British Library, the National Library of Scotland, Eton College Library, John Rylands Library (Manchester) and Cambridge University Library.

HIS PRINTING PRESS WAS INSPIRED BY ANCIENT OLIVE AND WINE PRESSES AND FEATURED A HANDLE THAT WOULD TURN A STRONG WOODEN SCREW.

PRINTING PRESS
Johannes Gutenburg created his game-changing printing press in 1436. His first print run was 180 copies of the *42 Line Bible* of which 12 complete copies printed on paper still exist.

WHEELBARROW

MACHINE NUMBER 016

A wheelbarrow combines the properties of both a wheel and a lever, which, in technical terms, makes it a "complex" machine – two or more simple machines working together. It enables you to move heavy loads easily. That's because, if you've got the load in the right place, you're really only lifting a small part of the total load. It's a genius invention and one of those devices that has changed very little, except for the materials used to construct it.

The earliest evidence for a wheelbarrow comes from China in the 2nd century CE. It's a mural on the outside of a tomb in Chengdu which dates to 118 CE and shows a man pushing a wheelbarrow with a particularly large wheel. This mural was created during the Han Dynasty – a 400-year period of remarkable inventions and innovation that included the suspension bridge, adjustable wrench, rudder and paper, to name just a few. Ancient Chinese wheelbarrows actually came in two forms – the type

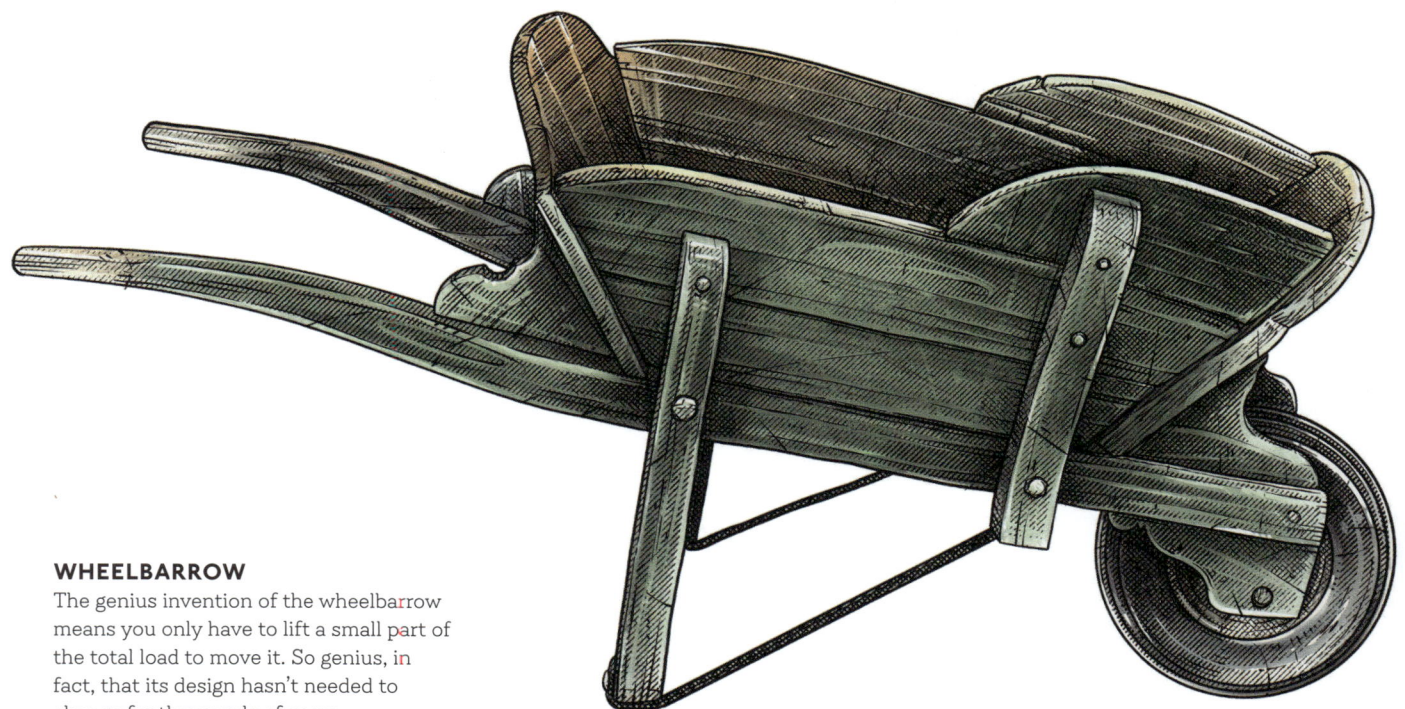

WHEELBARROW
The genius invention of the wheelbarrow means you only have to lift a small part of the total load to move it. So genius, in fact, that its design hasn't needed to change for thousands of years.

we're all familiar with, but also the one with a centrally mounted wheel, where the load is carried pannier-style on both sides. It's likely that both types of wheelbarrow had actually been around for centuries in China before the 2nd century CE. And that's even more incredible given that the wheelbarrow was completely unknown in Europe until around 1200 CE.

I spent a surprising amount of my youth pushing wheelbarrows around with three or four of my friends from secondary school. And that's because we spent every hour outside of school together making courses for our BMX bikes in the local woods, and this meant making a lot of steep-sided, triangular jumps by hand. And digging and lugging a lot of earth with a wheelbarrow. Almost all BMX courses are built by hand – many hands in fact – and the results can be seriously impressive. There's a camaraderie among the BMX crowd which I've always loved and a sense of etiquette, so if you damage a jump, it's your responsibility to make the repair. We used to pay to go to a BMX course called Chicksands, in Bedfordshire, which was made by local folks who also maintained the jumps. We'd make the long drive there in our old cars – me in my VW Beetle – and we'd ride around the woods for a bit on our bikes. One of us would always have a bag of tools because inevitably someone would get a puncture. Then we'd have a barbecue, often in the pouring rain. You can't get much more British than that. They were really happy days.

Looking back on it, I think we might have seemed a bit dodgy turning up in remote woodlands with wheelbarrows and shovels, but we were seriously committed to building something together that we loved. The amount of time we spent just digging mud was crazy, and it got crazier because the courses and jumps got more and more elaborate as we got older. We took a lot of pride in our work, and it got to the point where the tools we used – the spade and the wheelbarrow – were almost as important as the actual bikes. You wanted something you could rely on because we were using them so much. I remember buying a nice metal wheelbarrow from a car boot sale, with good hand grips and a solid wheel, so you know you're never going to get a puncture. Amazingly, despite my real fondness for them, I've never actually fixed one on *The Repair Shop*. Let's hope for one in Series 13.

THE WHEELBARROW WAS COMPLETELY UNKNOWN IN EUROPE UNTIL AROUND 1200 CE.

TREBUCHET

MACHINE 017

```
A trebuchet is a siege engine used to launch projectiles
over great distances. It's a little bit like a catapult,
but it offers huge advantages in terms of the weight of
the projectile that can be launched and the distance
that can be reached. Although it tends to be associated
with medieval warfare, it was actually developed in
China sometime between the 5th and 3rd centuries BCE.
```

The trebuchet features a beam on a raised fulcrum, and the beam is divided up into a short arm, from which a massive counterweight is hung, and a long arm, with a sling to hold the projectile. The short arm is raised up, the projectile is loaded at the other end and then the arm is released. Earlier versions of the trebuchet would have involved people pulling ropes attached to the short arm rather than using a counterweight.

The largest trebuchet in history had a fitting name for the terrifying medieval weapon it was: Warwolf. It was built on the orders of King Edward I (you might remember him as the really unpleasant chap in *Braveheart* who lobs someone out of a window) who'd laid siege to Stirling Castle in 1304 with hundreds of his soldiers. Stirling was the last Scottish stronghold of resistance to English rule after William Wallace had been defeated at the Battle of Falkirk in 1298 and was now on the run. King Edward spent four months trying to break down the walls of the castle with catapults, and smaller trebuchets firing lead balls and boulders, but nothing was working. So he got five master carpenters and 50 workmen to build an absolutely massive trebuchet. Just seeing this frightening weapon being built was enough to make the garrison in the castle offer to surrender. I'd have done the same. But Edward I had other plans, and refused to accept their surrender, ordering them back inside the castle and carrying on with the siege just so he could see the trebuchet in action. The trebuchet launched a 140kg (309lb) projectile which smashed a hole in the castle wall – and that was the end of that.

TREBUCHET
Although often associated with medieval warfare, the trebuchet was first developed in China between the 5th and 3rd centuries BCE.

CATAPULT

MACHINE 018

Weapons designed to hurl large stones towards enemies go back to the 7th century BCE and beyond in the Middle East and China, but we don't really know how they actually fired their projectiles.

Things get a bit clearer in Ancient Greek and Roman times, because the weapons were well described. In 399 BCE, Dionysius the Elder, the ruler of Syracuse, then a Greek colony on the island of Sicily, was making plans for a war with the city-state of Carthage. He wanted his engineers to come up with more advanced weapons, and they succeeded in devising a machine that drew and fired arrows with the help of twisted roped and/or animal sinew under tension. It was essentially a giant crossbow, and wasn't the first of its kind but one of the best known.

The more conventional-looking catapult developed from the need to fire projectiles faster and farther. And one of the most famous of these was the onager, which the Ancient Romans perfected. The onager featured a large wooden frame to which a smaller vertical frame was fixed, supported by two beams set at an angle of about 45°. A long wooden arm was attached at its base to a joist via very tightly twisted and plaited rope, which worked as a spring. Then you've got the conventional spoon-like container at the other end of the arm, which hurls the projectile. The arm is forced downwards to near horizontal using a geared winch. A locking gear stopped it from being released until it was time to fire. The stones the Romans used could weigh up to around 70kg (150lb) and were used to smash wall, turrets and ramparts. They were also used to fire all sorts of grim objects over castle walls, including the heads of captives and diseased carcasses. Lovely!

CATAPULT
The Romans developed the onager catapult that could fire bigger stones further. They also used it to lob the heads of prisoners and diseased carcasses over castle walls.

BELLOWS

A bellows is a device used to blow a jet of air out of a nozzle. It's essentially a hinged box or bag made of two flat boards connected by a band of flexible leather supported by wire rings.

A valve that can only open inwards in one of the two boards allows air to enter when you expand the bag or box (using handles positioned at the opposite end to the nozzle). The air shoots out through the nozzle when you bring the handles together. A bellows was commonly associated with a blacksmith's forge, to help speed up combustion, but they also became everyday items in homes with fireplaces.

One of the most famous bellows-makers was Thomas Linley, Sons & Co of Sheffield, one of the oldest companies in the city, with a history dating back to 1632. The company specialised in making industrial bellows, and this proved to be a vital part of the booming Sheffield steel industry in the 18th and 19th centuries when Sheffield was responsible for around 40 per cent of the total steel production in Europe.

At the Weald & Downland Living Museum, where we shoot *The Repair Shop*, there's a forge with a massive set of original bellows behind. It always takes two people to work the forge because one of them is just pumping the bellows pretty much the whole time. There's a handle sticking out made of cow horn, and it has become highly polished over time because it has been used by so many hands. Cow horn is actually a really good material to use because it's very strong and durable, and can be shaped easily. It's also traditionally used for the nozzle of bellows, which concentrates a jet of air to fan the furnace. Operating the bellows is a lot harder than it looks – it takes some serious strength to pump the bellows at the forge when they're full of air, I can tell you.

PENDULUM CLOCK

MACHINE NUMBER 020

Christiaan Huygens was a Dutch mathematician, physicist, engineer and astronomer who made huge breakthroughs in almost every field he worked in. Among them was discovering Titan, the largest moon of Saturn, with one of the telescopes he built. He also came up with groundbreaking theories on centrifugal force and light waves. Oh, and he invented the pendulum clock. This is one of those rare inventions where we know exactly when he invented it, to the day: 25 December 1655. He patented it the following year, and entrusted building his clock designs to notable clockmaker Salomon Coster from The Hague. One of Coster's earliest examples, from 1657, is in the Rijksmuseum Boerhaave in Leiden, the Netherlands.

Huygens wasn't the first to design a pendulum clock, though. That honour belongs to another scientific legend: Galileo Galilei, in 1642, the year of his death, and his son partially constructed it in 1649. It had been Galileo who realised that a pendulum takes the same time to complete one swing no matter how wide the arc is. What Huygens did was make the pendulum truly constant by developing a pivot that made the suspended "bob" swing in a very specific type of circular arc (technically it's called a cycloid – a curve traced by a point on the circumference of a circle that rolls along a straight line. And that's why I went with *circular arc* instead). The pendulum is connected to a toothed wheel. Every time the pendulum swings, an anchor (called an escapement) releases one tooth. It's this mechanism that makes the *tick-tock* sound. This escapement mechanism gives the pendulum a little injection of energy, so that it both transfers enough energy and overcomes friction and drag. The escape wheel is also connected via a gear train to a weight. Each second, the weight falls a bit and this moves the seconds hand. This is attached to another gear train

THEY WERE ONLY NAMED GRANDFATHER CLOCKS AFTER THE POPULAR 1876 SONG "MY GRANDFATHER'S CLOCK" PENNED BY AMERICAN SONGWRITER HENRY CLAY WORK.

that also moves the minutes hand after the seconds hand has moved 60 times. And the minute hand moves the hour hand at 1/60th of its speed.

The pendulum clock was by far the most accurate clock invented to that point. It became the global standard for timekeeping for over 270 years, until the quartz clock came along in 1927. Pendulum clocks were initially handmade by craftspeople, and were therefore beautiful but expensive objects, but during the Industrial Revolution, factories began producing clock parts in large quantities, and so they became more affordable.

If you're wondering where the grandfather clock fits in, they were originally known as long-case clocks. They featured long pendulums and weights suspended by cables or chains and are often stunning pieces of decorative art. The earliest forms of them were more functional than beautiful and were made around 1670, with some very famous names emerging, like William Clement, Edward East and Ahasuerus Fromanteel. They were only named grandfather clocks after the popular 1876 song "My Grandfather's Clock" penned by American songwriter Henry Clay Work, after he saw one in the George Hotel, in Piercebridge, County Durham.

PENDULUM CLOCK
For 270 years pendulum clocks were the global standard for time-keeping. Their accuracy is down to the very specific circular arc – technically a cycloid – that the pendulum swings along.

MECHANICAL CALCULATOR

MACHINE NUMBER 021

Blaise Pascal was a child prodigy who came up with, and constructed, a mechanical calculator in 1642, aged just 18. It wasn't like he did it for financial gain either – he invented it because his dad was a tax collector and he felt sorry for the poor guy, spending months calculating, and recalculating, sums by hand.

So many tedious household tasks have led people to invent ingenious machines. Pascal wasn't adapting something that had already been made – he thought up a novel idea and then made it come to life. It's a really rare blend of qualities, coming up with a vision and then having the self-belief, commitment and stubbornness to see it through. It takes a lifetime for some folks, but Pascal's achievement, when he was just 18 years old, beggars belief.

Pascal's machine performed both addition and subtraction, and became known as the Pascaline. It had six metal discs that looked like spoked wheels. The one on the far right was for single units, the one next to it for tens, then hundreds, thousands, tens of thousands, and hundreds of thousands, and above each disc was a display "window". You entered a number by turning a stylus on the relevant dial, then you entered another number and the machine would show the sum of the two. Behind each wheel was a horizontal gear, which rotated a vertical gear attached to a horizontal axle. When a number needed to be "carried over", a weighted ratchet between the gears nudged the next gear around a notch. It was a very clever machine for its time, and eight of them still survive, four of them owned by the Musée des Arts et Métiers in Paris.

In 1812, English mathematician, inventor and mechanical engineer Charles Babbage came up with an idea for a machine that could mechanically calculate mathematical tables, mainly to help with navigation. At the time, sailors, engineers, astronomers, architects and many others used printed tables to find the answer to complex arithmetic. The problem was, everyone makes mistakes in books, and when you're relying on accurate calculations for, say, a big engineering project, an error could be really costly. Babbage wanted something completely reliable, so he got to work.

He called his design the "Difference Engine No. 1", and persuaded the government to help fund it. Master toolmaker Joseph Clement was the person actually building it, and finished a section of it in 1832. It was basically a working model of the machine and involved approximately 2,000 parts – around one-seventh of the whole machine. The entire machine would have filled an entire room.

The machine wasn't designed to work like a modern calculator, working out arithmetic that you could input into the machine, and it wasn't capable of doing multiplication or division – it performed simple repeated additions and was designed to print the results. The working model featured columns of interacting gear wheels with dials engraved with the numbers nought to nine. It was all held together in a brass framework and operated by a crank at the top. It's now in the Science Museum in London. Not only is it a mechanical masterpiece – it's also a thing of beauty.

SURVEYOR'S WHEEL

MACHINE NUMBER 022

A surveyor's wheel, also called a measuring wheel, viameter, waywiser and perambulator (among many other names), is a device used to measure distances. This ancient machine goes back to at least the 3rd century BCE in the Roman Empire, and the 1st century CE in Ancient China. It's essentially a wheel that rolls along the floor and is attached to a handle.

Many measuring wheels were made so that the circumference of the wheel was a specific measurement, say 1m (3ft), so that each time you completed one revolution of the wheel, you knew you'd travelled 1m (3ft). Where it got slightly more complicated was recording the number of revolutions of the wheel, because, as anyone who's ever tried it will tell you, you do tend to get distracted when counting up to high numbers from zero. The simplest step up from manually counting was connecting the wheel to a moveable device that produced an audible click after one complete revolution. But again, that had its disadvantages, so mechanical devices involving gears (see page 18) were connected to the wheel itself, with a dial registering each revolution.

Measuring wheels became more common instruments in the 17th century when distances were being routinely measured to establish land boundaries, and in road building. Some of the surveyor's wheels that were made during this time, and into the 19th century, are beautiful objects. There's one fascinating measuring wheel up in Knaresborough, North Yorkshire, which was specially designed for a chap called John Metcalf (better known as "Blind Jack"), who'd gone blind after catching smallpox when he was six. Blind Jack became a legend in that part of the country, saving a soldier from drowning in the River Nidd when he was just a boy. He got a job as the resident fiddler in a well-known Harrogate hotel, later working as a local tour guide. In 1745, aged 28, he joined a militia put together to help fight Bonnie Prince Charlie (Charles Edward Stuart, the grandson of King James II, who claimed the English, Scottish and Irish throne from 1766). Blind Jack was present at the Battle of Culloden – the final defeat of Bonnie Prince Charlie's Jacobite forces in 1745. In the 1750s, the

MEASURING WHEELS BECAME MORE COMMON … WHEN DISTANCES WERE ROUTINELY MEASURED TO ESTABLISH LAND BOUNDARIES, AND IN ROAD BUILDING.

forward-thinking Blind Jack became the owner of a stagecoach company. Seeing the terrible conditions of the roads, he got involved in road construction, even managing to build roads across marshes and bogs. In all, he was responsible for building around 290km (180 miles) of new roads across Yorkshire and Lancashire. And he used his trusty custom-made measuring wheel to measure out distances.

Edward Troughton was a renowned scientific instrument maker, based at 136 Fleet Street in London, from the late 1770s, in partnership with his brother, John. Their shop, which they ran with their uncle John, was named the "Sign of the Orrery", which sounds like something out of a Charles Dickens novel. Edward Troughton (whose company became Troughton & Simms in 1826) constructed stunning surveyor's wheels made of pine with brass gauges and intricate clock-like dials. They go for an absolute fortune now, but are very simple devices for Troughton's standards, given that he designed some hugely complex instruments, including the equatorial telescope at Armagh Observatory, thought to be the oldest telescope still housed in its original dome.

SURVEYOR'S WHEEL
The circumference of the wheel measured a certain distance, while the dial recorded how many revolutions it had made. The machine was essential when road building took off in the 17th century.

SUBMARINE

MACHINE NUMBER 023

In 1578, the English writer, mathematician, volunteer gunner and innkeeper William Bourne came up with an idea for a wooden boat that could travel underwater. He realised that if you make a boat heavier than the weight of the water which the boat displaces, it will sink. He suggested enveloping the boat in waterproof leather, using a screw mechanism to move the boat up and down and expanding and contracting the interior of the boat to change the volume.

As far as we know, he didn't draw up his idea or attempt to make one, but the principle may well have worked if he had done. Dutch inventor (and former engraver and glassworker) Cornelius Drebbel did build a "diving boat" that submerged into the River Thames in 1620. It was basically an upturned wooden rowing boat covered in greased leather and propelled by oars. It had a watertight hatch in the middle and inflatable pigskin bladders were thought to control the depth. Pipes attached to two of the bladders supplied the air for the 12 poor guys rowing the makeshift submersible, but it worked, keeping them at around 4.5m (15ft) under all the way from Westminster to Greenwich. It's hard to understand who on earth actually volunteered to get into this crazy contraption. People were a lot braver back then.

Things moved along a lot towards the end of the 18th century. In 1776, in the second year of the American War of Independence/Revolutionary War, American inventor David Bushnell came up with a one-person submersible to attach mines to British warships. The *Turtle* was 2.4m (8ft) long, shaped like an egg, and used a foot-operated valve to let in enough water to submerge it, and two treadles to operate the propellers that drove it forward and up and down. George Washington, the Commander in Chief of the Army and a big fan of Bushnell's, set his sights on the British flagship – *HMS Eagle*, which was in New York harbour. But Bushnell wasn't strong enough to operate the difficult controls, so a sergeant called Ezra Lee was chosen for the task. The plan was to use a hand-powered drill to make a hole in the hull and attach a large keg of gunpowder before getting away as quickly as possible. It sounds like something out of a *Looney Tunes* episode with Daffy Duck trying to get one up on Bugs Bunny.

Everything went to plan until the drill didn't make any headway into the hull, so he ended up resurfacing and making a hasty retreat. He detached the mine, which later exploded, but it was nowhere near the *Eagle* – or the *Turtle* for that matter. The submersible was eventually lost when the sloop it was traveling aboard was sunk by the British.

The *Nautilus* was the next serious attempt to develop a submarine, conceived by the American artist and inventor Robert Fulton, between 1793 and 1797. Fulton was living in Paris, France and offered to construct a submersible to attack the British, then at war with the French. The submarine was shaped like a cigar, constructed of copper plates on iron ribs. It featured a 4.7-m/15.5-ft-long hollow-iron ballast tank, which allowed water in through a valve to make the vessel sink. A screw propeller, which two men operated by cranks that drove gears, could make the *Nautilus* reach around 2 knots (2.3 mph). It was steered via a long-handled crank connected to the rudder spindle via an arrangement of cogs. Like Bushnell's creation, the *Nautilus* was designed to carry a mine to attach to an enemy ship. It performed amazingly well in tests, diving down to 7.6m (25ft) – which had never been achieved before – and even destroying a stationary 12m (40ft) sloop with its naval mine. But try as he might, Fulton could not convince Napoleon Bonaparte that the submarine was worth investing in. The British thought otherwise, and offered to fund his work on a second *Nautilus* to use against the French, but not long after, Admiral Nelson scored such a decisive victory over the French at the Battle of Trafalgar in 1805 that his services weren't required. Like so many inventors before and after him, timing wasn't on his side.

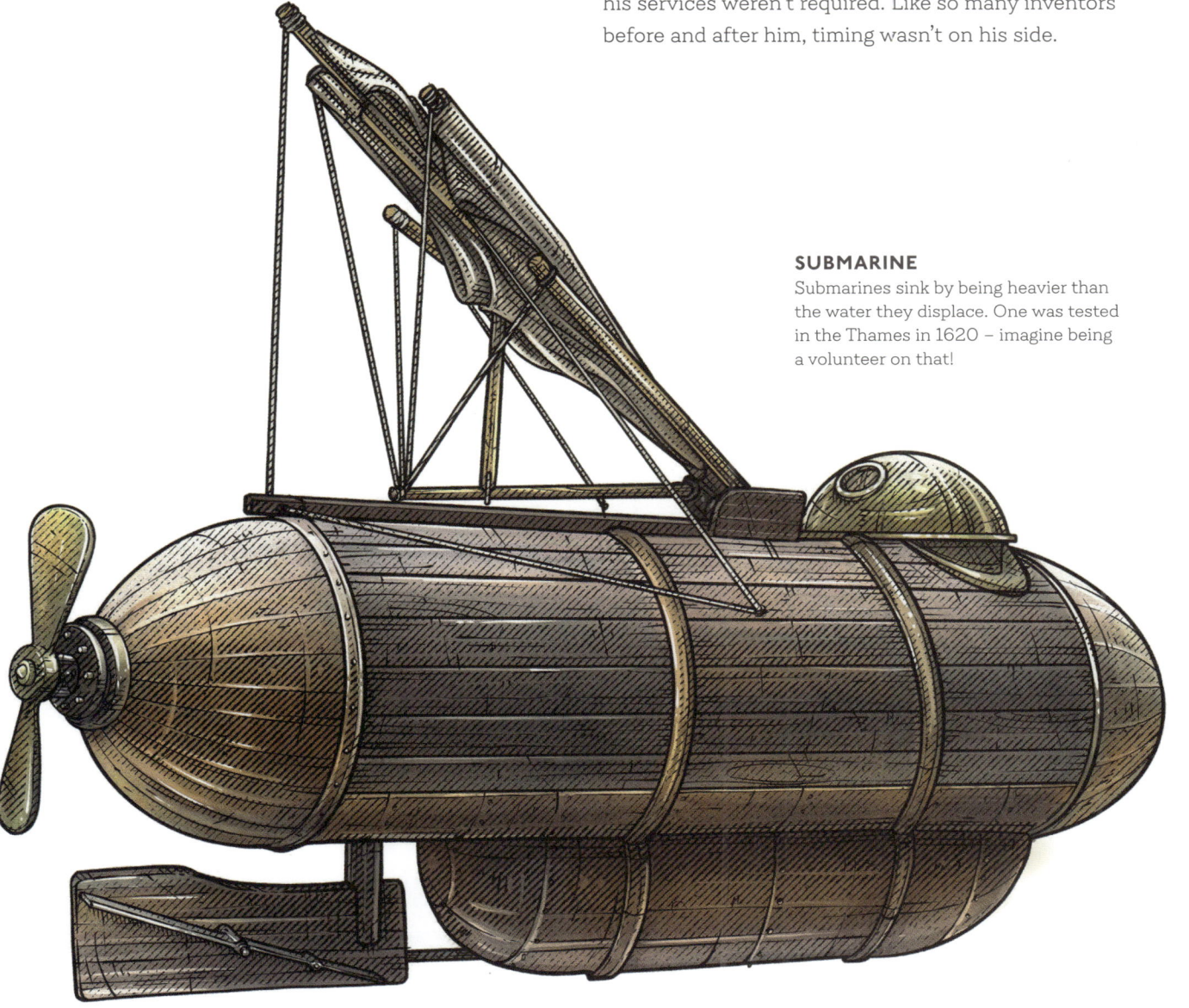

SUBMARINE
Submarines sink by being heavier than the water they displace. One was tested in the Thames in 1620 – imagine being a volunteer on that!

CHAPTER 2

FOOD AND DRINK

MACHINE NUMBER	MACHINE NAME	PAGE NUMBER
024	BUTTER CHURN	048
025	COFFEE GRINDER	051
026	CORKSCREW	054
027	EGGBEATER	057
028	MEAT GRINDER	058
029	CHEESE PRESS	060
030	PASTA MACHINE	062
031	BEER ENGINE	064
032	PEPPER GRINDER	065
033	SODA SIPHON	066
034	NUTCRACKER	068
035	MECHANICAL SCALES	070
036	CAN OPENER	072

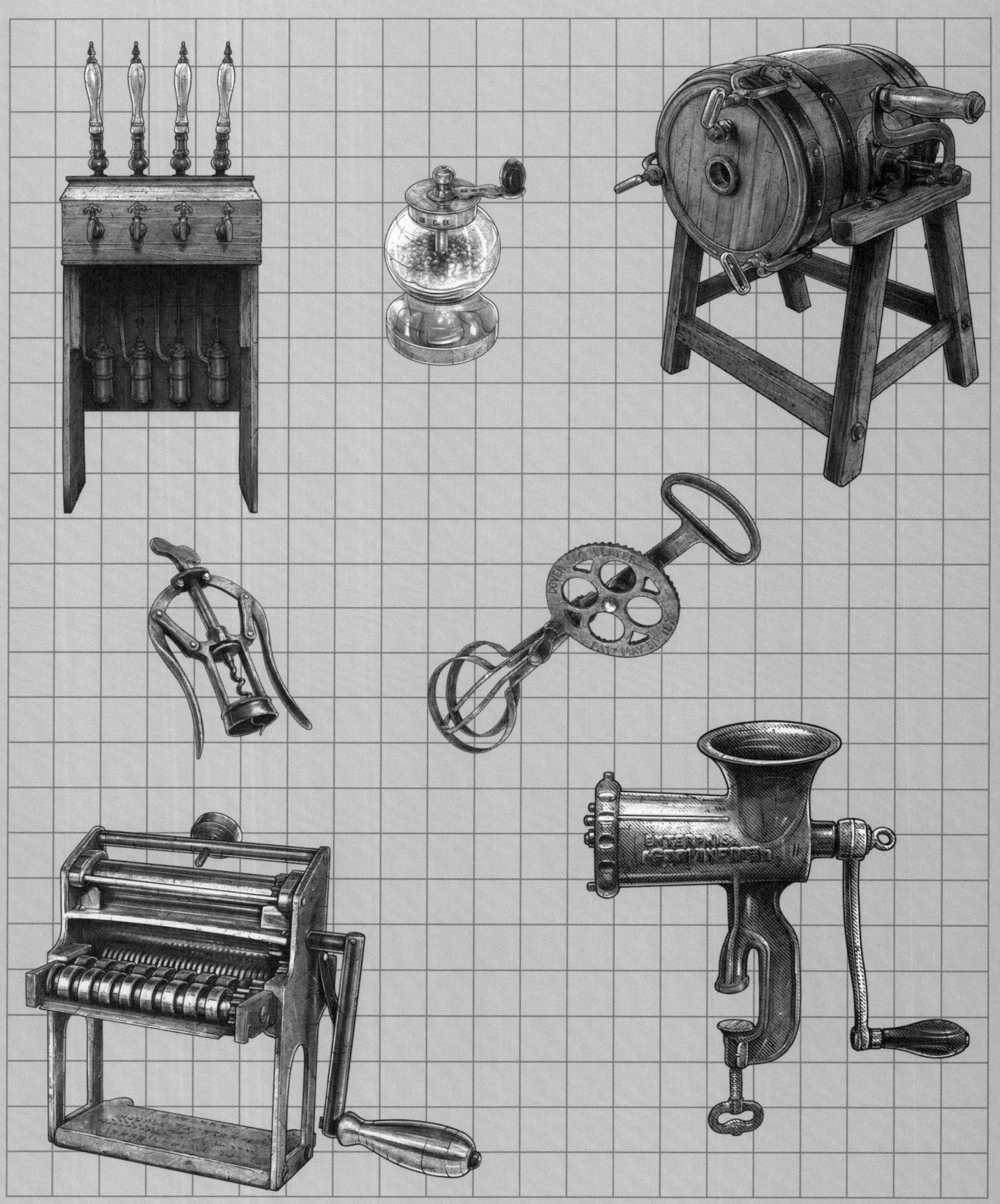

BUTTER CHURN

MACHINE NUMBER 024

A butter churn is a mechanical device that changes cream into butter by separating the fat from the buttermilk. As for how old they are, it's hard to say exactly, but we do know that early butter churns existed in the Copper Age, approximately 3200 BCE–2300 BCE.

There's a pottery butter churn in the Israel Museum in Jerusalem with handles on each side from which ropes were slung, so that you could manually rock the churn back and forth. That one dates from the Copper Age. But there were probably earlier ones just made of animal skin and manually shaken – which would have been some physical activity, I can tell you.

In the Middle Ages, there was one type of butter churn – the dash churn. It was a simple enough device, basically a barrel with a lid through which you plunged your stick (called a dash) and moved it up and down to whip the cream up. You'd look like Popeye after using this for just a few weeks, I bet, because manually churning butter takes bloody ages! We're talking several hours of hard labour. And what's incredible is that this was happening in most homes. The dash churn changed very little over time and it was the most common butter churn from the Middle Ages all the way to the Industrial Revolution.

The barrel churn was a technological step forward, designed to make churning butter less back-breaking. The churn was essentially a barrel sitting on wooden supports or legs and a crank arm attached to the barrel, which the user would turn, moving either the entire barrel itself (vertically or horizontally) or, later, the contents of the barrel via a paddle. Some of the design for barrel churns were pretty unbelievable, including ones that looked like baby's cradles, rocking horses and rocking chairs, but they were clever in that they utilised momentum to cut down on the amount of arm-work needed. The arrival of the paddle churn in the 1800s was a big deal. Although it manipulated the contents of the churn in a similar way to the later barrel churns, it could be a much smaller contraption, because instead of moving the entire churn, it manipulated the contents via a crank attached to a paddle inside the churn.

One English family-run company that made a name for itself with the quality of its butter churns was Hathaway in Chippenham, Wiltshire. It started with George Hathaway, who became an apprentice to a cooper, moved to Chippenham and established a butter churn-making business. The machine with which he became associated was the "over and over" churn, a stand-mounted barrel that featured a small window so you could see the fruits of your labour. After about half an hour, your butter was ready. Its "over and over" churn has a cast plate featuring the company name, the patent and sometimes even instructions on the right temperature you need to make the best butter – 13°C (55°F) in summer and 14°C (58°F) in winter, in case you're wondering. Hathaway made a name for itself and began advertising in London trade catalogues, winning over 30 medals at agricultural fairs for the quality of its churns. There's an especially fine one in the Chippenham Museum.

I fixed a butter churn on the first episode of Series 3 of *The Repair Shop*, back in 2018. I worked with my pal Alastair Simms on it, one of the last remaining master

BUTTER CHURN
Butter churning got less and less back-breaking with every technological innovation. The paddle churn manipulated the contents within the barrel rather than turning the whole barrel.

coopers in England. I remember fixing the lid and the clasp to hold it on and making a rubber gasket to make sure it didn't leak. It had the instructions for how to use the churn cast into a metal plate and I remember cleaning and colouring in all the letters as part of the repair. It read, "'Champion' Churn – Used by the Champion Butter Makers of England & Scotland, Dairy Supply Co Ltd, London, Edinburgh & Cork". You can actually still see where the company used to be based, because the lettering is set around the image of a magnificent building, dating from 1888, on the corner of Coptic Street and Little Russell Street in London, just round the corner from the British Museum. The company was founded by George Barham, who is sometimes referred to as the "father of the dairying industry in the UK", as a separate arm of the Express County Milk Supply Company. Barham famously came up with the idea of transporting milk by train from rural farms, which couldn't have come at a better time because a cattle plague in 1865 sadly wiped out most of London's cows. By 1885, his company was delivering more than 136,000 litres (30,000 gallons) of milk to London every night. He also invented the milk churn (the tall cylindrical container that made it easy to transport large quantities of milk) and his company was the first to use glass milk bottles. He campaigned for cleaner milk and his company became one of the first to pasteurise milk. Barham even supplied milk to Queen Victoria.

I still remember what we call "the reveal" on the show for the butter churn. The producers didn't want me to test the churn until that point, which was quite nerve-wracking because it could have gone horribly wrong if there was a leak of any kind, the way the churn flips around so quickly. But thankfully, it all went well – we'd bought in 10 litres (2.2 gallons) of cream from Goodwood Dairy and poured it into the churn, hearing it sloshing around as we turned the crank. It takes a good 20 minutes or so of spinning it until the sloshing sound suddenly changes into a thud, and that's when you know the butter's ready. It's an amazing process.

We actually fixed another butter churn on *The Repair Shop* for Series 12 – but this one was a little one called a Dazey Churn. It's like a glass Kilner jar, small and transparent with a screw-on lid and wooden paddles and a geared crank handle on top. It's designed to just make a little block of butter. One of the runners on the show told me that making butter at home has become a massive thing on TikTok. I felt so old! But people are becoming so aware of where things come from and how things are made, and home-made butter has become a thing. Kilner themselves have got in on the act, producing butter churns to use at home. And that makes me very happy – I love hearing that a heritage craft is undergoing a revival.

IT TAKES A GOOD 20 MINUTES OR SO OF SPINNING IT UNTIL THE SLOSHING SOUND SUDDENLY CHANGES INTO A THUD, AND THAT'S WHEN YOU KNOW THE BUTTER'S READY.

COFFEE GRINDER

We don't know exactly when people started roasting and brewing coffee, but we do know it was happening in Yemen in the 15th century – although the coffee plants were almost certainly taken from Ethiopia and cultivated in southern Arabia.

The popular legend goes that around 800 CE, a goatherd in what was then Abyssinia, now Ethiopia, noticed some bizarre, excitable behaviour among his flock. Called Kaldi, he saw that they were feeding on some type of berry and tried it himself. (What would possess anyone to do that is beyond me, but I suppose that's the origin of discovering every potentially lethal foodstuff; someone's got to give it a go.) So Kaldi gobbles a few berries, finds himself full of beans and proceeds to tell everyone about it. It's a good story, which means it's probably rubbish. But who cares?

The first coffee grinder would have been the pestle and mortar, and after that, the spice grinder, but they just weren't up to the job to get the fineness needed for coffee. It was between 1400 and 1500 that the Turkish pocket-cylinder coffee mill first appeared. This was a simple machine, using a handle to rotate one grinding surface against another. Fast-forward to the 18th century, and mills were being imported to western Europe from Turkey and were therefore rare and expensive. So up stepped a (now-famous) French family business: Peugeot. Long before Peugeot made bicycles and cars, they were making top-quality coffee grinders. The family business dates back to the early 19th century and the commune of Valentigney in eastern France near the border with Switzerland. The Peugeot family had a grain mill, which two brothers – Jean-Pierre and Jean-Frédéric – transformed into a steel foundry in 1805. They started off making saws, but they really made a name for themselves only after they developed a new process of cold-rolling steel by tempering the steel before flattening it under pressure. They used the rolled steel to make a number of hand tools as well as supplying steel to the watchmaking industry.

In 1839, their saws became world-famous after they developed a laminating process to thin the steel used for their saw blades, which made cutting

much quicker. But what they didn't have was an emblem which would make their tools instantly recognisable at a time when not everyone could read or write. They wanted something that conveyed the strength and power of their products, so they came up with the lion standing on an arrow. It summed up their saw blades: strong, fast, flexible, and with a serious bite.

In 1840, they branched out by producing their "R model" coffee mill, made from steel and wood, which was available in ten different sizes. The grinding mechanism featured a crank that rotated an abrasive paddle against a fixed, spherical toothed surface. You could adjust the height of the paddle so the two surfaces were closer together (to get a finer grind) or further apart (for a coarser one). Underneath was a neat wooden drawer which collected the coffee. Whatever the height of the paddle, the machine gave an even grind without heating the beans, which would affect the taste of the coffee when it was brewed. For all the non-coffee drinkers out there wondering why freshly grinding coffee beans is so important, it's because that's the way you maximise the aroma you get from the beans. The reason Peugeot mills were so good, and their mechanisms have stood the test of time, is because the steel was so strong and durable.

ORRIE McCOMBS

COFFEE GRINDER
The art of a good coffee grinder is to produce an even grind without warming up the beans, which would change the taste of the coffee when it brewed.

CORKSCREW

MACHINE 026

I realise as I'm writing this that a lot of machines we use every day were inspired by tools and machines of war. It turns out (there'll be a few of these jokes) that the corkscrew was most probably inspired by an early 17th-century device known as a gun worm.

The gun worm was a 20–30cm (8–12in) piece of iron, tapering towards a coiled tip with one or two sharp points; it was attached to a long wooden pole and turned a few times to retrieve any debris (the remnants of a spent powder bag, for example) from the barrel of a cannon or musket before you loaded it again.

It isn't exactly known when the first corkscrew appeared, but the earliest written reference to one appears in 1681 in a London catalogue of the *Natural and Artificial Rarities Belonging to the Royal Society* which mentions "a steel worme used for the drawing of Corks out of Bottles". The context of the entry suggests it was a really common item at the time. And this makes sense, because bottle designs had changed in the middle of the 17th century to feature slimmer necks. Having said that, though, Guinness World Records has the earliest documented corkscrew – a French corkscrew with a cage-like central section – dating to 1685.

One thing we do know for sure is that the first patent for a corkscrew was granted to Reverend Samuel Henshall in 1795. Henshall was the Rector of London's Bow Church, where he's buried. He had come up with a button-shaped concave disc at the top of the worm (connected to a cigar-shaped wooden handle), which made it simpler to break the seal between the bottle and the cork, stopped the worm from going too far into the cork and prevented the cork from breaking up. In 2009, a plaque was installed at Bow Church to commemorate Henshall's invention. The Reverend Michael Peet, then Rector of Bow Church, was said to "be pleased to note that such a noble instrument has ties to this Church and further suggests that the inventor of the screw cap should be relegated to Hell". Oddly enough, my first workshop was opposite Bow Church, so I walked past it every single day for years. I'd just quit my job at Rankin Studios as a set designer and it was my first solo workshop. It was such a big move for me at the time – it was an exciting but terrifying leap into the unknown, starting a business up from nothing, but it was what I always wanted. I remember opening up the metal shutters and seeing an empty industrial unit, with water leaking in from the knackered roof and ivy growing inside. Most people go down the route of sharing a communal workshop, and to be fair, that would have been a much more sensible way of doing it in some ways, but I like knowing that everything I've left somewhere is still there the next morning. In my new workshop, after I sorted out the leaks and the ivy, I started making fairground lights – light-up letters, shooting star designs, and lots of other shapes that had become massively popular at the time. I'd shape these designs, distress them, put bulbs in them, before sign-writing and painting them. I couldn't make them quickly enough, and was probably selling them far too cheaply, but I was making a bit of a profit and it was enough to cover the rent, so it kept me going. I went to a classic boot sale in East London and brought along a few of the light-up shooting star designs I'd been working on, and I ended up meeting the owner of a big

CORKSCREWS
An advert showing various different types of corkscrew.

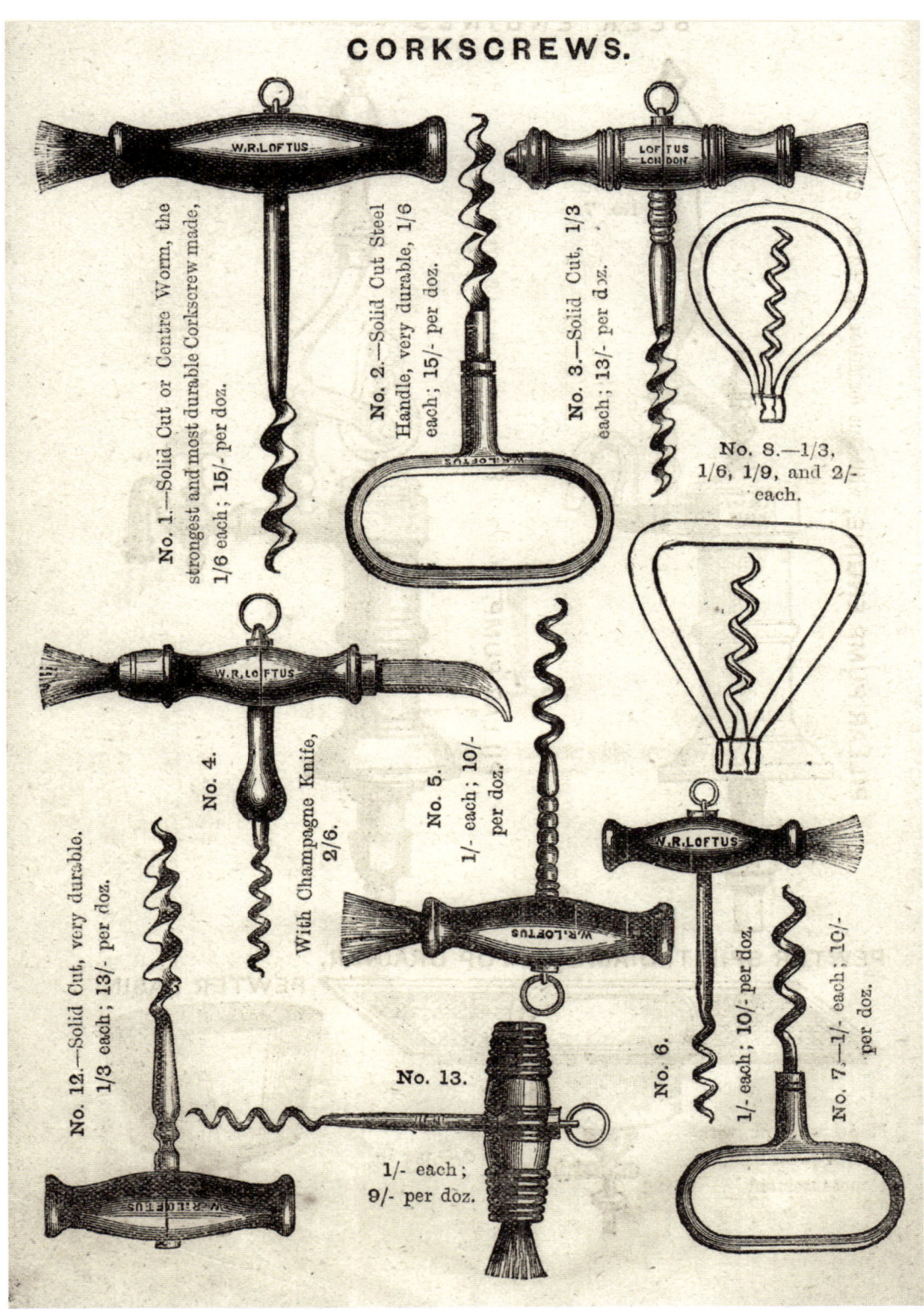

DOUBLE LEVER CORKSCREW
The A1 Heeley Double Lever corkscrew was based on a design patented in 1880, but Heeley made the small but transformational modification of connecting the levers to a collar.

events company, who'd just wandered past my car boot. He saw the stars, gave me a card and asked me to get in touch. I met him, we got on really well and I worked for him on all sorts of lighting designs for events until I started on *The Repair Shop*. And that journey all began by Bow Church, although I had no idea about its historic connection to the corkscrew until I did some delving for this book.

The winged corkscrew – the one with two levers that rise up together as the worm screws into the cork – traces back to the first double-lever corkscrew, patented by Englishman William Burton Baker in 1880. It was made by James Heeley & Sons in Birmingham, UK, after Heeley decided that the design could use modification to make it smoother and less liable to break. So he developed the A1 Heeley Double Lever, which had the levers connected to a collar that slid on the shaft. He had it patented in 1888, although only in the UK. It was that design that really stood the test of time and provided the basis for the many variations that followed around the world.

I love that sommeliers choose the Swiss-Army-knife-style basic corkscrew with no flip-out bit at the end and no fancy levering opening. It's the same with me and my pliers and screwdrivers. I know them, I trust them, I look after them, and they work really nicely. Yes, there are probably technically better ones, but that makes no difference to me. I bet it's the same for a sommelier. Everyone has their tools and machines that they swear by.

I LOVE THAT SOMMELIERS CHOOSE THE SWISS-ARMY-KNIFE-STYLE BASIC CORKSCREW WITH NO FLIP-OUT BIT AT THE END AND NO FANCY LEVERING OPENING.

EGGBEATER

The first eggbeater was patented in the US in 1856 by tinner Ralph Collier of Baltimore, but it didn't look anything like the hand-held device now common in many home kitchens.

MACHINE NUMBER 027

A lidded box-like contraption mounted on legs, it was more of an industrial model for use in hotels and restaurants. In 1857, London inventor E. P. Griffiths came up with a rotary eggbeater that was a lot like a butter churn, with a crank-turned wheel mounted over a barrel. Griffiths' invention claimed that one complete turn of the handle "gave 288 strokes". If he's right, that is quite the labour saver.

The first hand-turned eggbeater that looks similar to one you might have in your kitchen drawer – with its hand-cranked drive wheel that turns the bevel pinion that spins the beaters – was patented in 1859 by J. F. and E. P. Monroe, although it was attached to a table via a screw clamp. Then followed a flurry of egg-beating inventions, many of which (including the Monroes' one) were swallowed up by the Dover Stamping Company. And this company's success was such that other US companies cheekily marketed their devices as "Dovers" as well. Dover was awarded a UK patent and began to manufacture its products there, but its eggbeaters didn't stir up the market in the same way as they had in the US. The electric-powered eggbeater followed in the 1880s, but despite the success of the electric models, many people still favour the good old-fashioned device that has changed so little in 160 years.

THE DOVER EGGBEATER
This staple of kitchen drawers first appeared in 1859 and its design has hardly changed since. It was so successful that other companies used to market their products as Dovers.

MEAT GRINDER

Machine Number 028

```
Karl Drais was a prolific German inventor, who we'll talk
about later on in the book because he came up with the
earliest form of a bicycle (see page 222). He also
developed a foot-driven railway cart and a stenography
machine. He is thought to be the first person to create
a grinder specifically for mincing meat.
```

Meat was fed through a funnel and hand-cranked via a screw conveyer into a special metal plate containing holes that forced the meat into long strands. It wasn't until the Philadelphia Centennial Exposition in 1876, the first official World's Fair to take place in the US, that the meat grinder really took off. Although, it might not have been what everyone was talking about at the fair, seeing as this was also the first time that Alexander Graham Bell's telephone was shown to the public. And Thomas Edison's automatic telegraph system was on display too, as was the Remington No. 1 Typewriter and Heinz Ketchup. Quite a collection. What I would do to go back in time and see what else was at that fair! The meat grinder was also a game-changer, though, not only because it was a time saver at home and a fantastic tool to help butchers offload less than desirable cuts of beef. It also helped to shape a cultural phenomenon: fast food.

Enterprise Manufacturing Company was one of the Philadelphia-based meat-grinder producers, although they also produced a number of household products, including a cherry-pitting machine, a coffee mill, sausage stuffers and graters, and some more unusual items, including a smoked-beef shaver, a meat-juice extractor, and a fruit, wine and jelly press. Founded in 1864, they were in business until 1956. Another name from the 19th century you come across, this one from the other side of the Atlantic, is Spong & Co. They were established in 1856 in London and made all sorts of domestic machinery for the emerging Victorian middle class who didn't have the cash to splash on servants. But it wasn't just the middle class who had a Spong in the kitchen. Queen Victoria did as well. They won all sorts of major awards for the quality of their kitchen products, including a gold medal at the Paris Exhibition of 1892. They were still producing mechanical kitchen products until the 1970s and were eventually bought by Salter Housewares Limited, who are still around.

IT ALSO HELPED TO SHAPE A CULTURAL PHENOMENON: FAST FOOD.

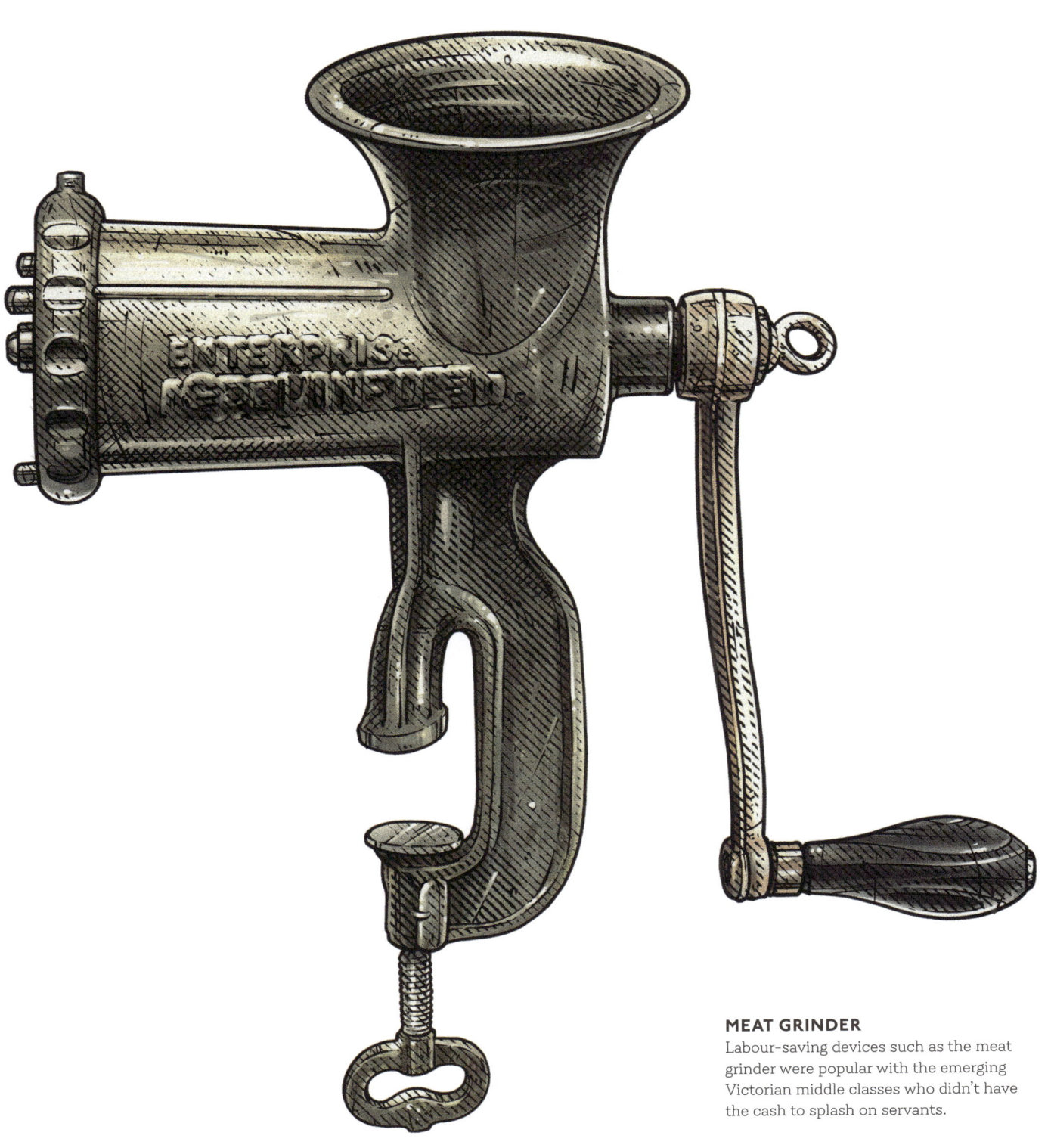

MEAT GRINDER
Labour-saving devices such as the meat grinder were popular with the emerging Victorian middle classes who didn't have the cash to splash on servants.

CHEESE PRESS

MACHINE 029 NUMBER

The cheese press is an ancient invention, going back to at least Ancient Egypt, but it was the Romans we've got to thank for introducing cheesemaking to the UK and across Europe.

Cheese is made (nowadays) by adding an enzyme called rennet to milk, which makes the milk curdle in a controlled environment. You don't have to use rennet, though – any edible acidic substance, like lemon juice, vinegar or sour wine will do, which are all among the things Romans used instead of rennet. Or you could just wait for the milk to turn sour, but probably best not to do that at home.

What you end up with when you've deliberately curdled your milk is a thin, watery liquid, called whey, and solid masses of milk proteins called casein, or more commonly curds, which is the part that becomes cheese. Where the cheese press comes in is not only to separate the curds from the whey but to force the curds into a shape. How much you press the cheese, and what you add in afterwards, is what determines the finished style of the cheese. Roman examples of cheese presses are circular ceramic or pottery bowls with one or two internal raised circular rings and perforations to drain away the whey.

Cheese was a fantastic solution to the problem of having milk that you didn't know what to do with. Monasteries became cheesemaking centres across Europe because they kept cows, sheep and goats and so had plentiful supplies of milk, plus it was a great way to make a bit of cash for monastery upkeep. Regions developed their own styles of cheese. Places such as France and Italy produced more types of cheese because of their wide regional variation in climate and geography. In the UK, we know that monks were making cheddar from the 12th century, and an accounting document from the time of King Henry II of England tells us that he bought 4,645kg (10,240lb) of cheddar in 1170, so he was definitely a fan. Although, possibly controversially, the cheddar wasn't from anywhere near the town of Cheddar in the southwest – it was from Byland Abbey in Yorkshire.

In medieval times, cheese presses would have been wooden contraptions, but the Industrial Revolution ushered in a period of innovation in cheesemaking.

HOW MUCH YOU PRESS THE CHEESE, AND WHAT YOU ADD IN AFTERWARDS, IS WHAT DETERMINES THE FINISHED STYLE OF THE CHEESE.

Large cast-iron cheese presses could now be manufactured to produce cheese on an industrial scale. Thomas Corbett of Shrewsbury founded the Perseverance Ironworks in 1865 and produced agricultural machinery that he exported around the world. But he was also a cheese press specialist. If you ever come across the name Thomas Corbett on a cast-iron piece of machinery, you know that you're looking at something made by someone at the top of their game. I saw an example recently from 1880 with an acorn-shaped finial and a brass pressure gauge below, reading in hundredweights. The press has three circular cast-iron stands to press cheese wheels, and two of the platforms move up or down and are operated by a large turning capstan. It's no surprise that Corbett won over 1,000 different awards for his machinery, went on three world tours to promote his wares, and even met British, Belgian and Dutch royalty.

CHEESE PRESS
It is what you add in to the curds, and how much you press them, that determines the finished style of the cheese.

PASTA MACHINE

MACHINE NUMBER 030

The first depiction of a pasta press dates back to 1615, but it wasn't until 1767 that Frenchman Paul-Jacques Malouin gave us a proper visual demonstration of how a machine that made vermicelli worked.

It was a contraption involving a large wooden screw that was cranked around, forcing dough that had been placed in a special chamber through a perforated plate (or die) made of bronze, creating the desired pasta shape. If you wanted to make macaroni, for example, the perforations would have a steel pin in the centre to create the tube.

Thomas Jefferson, who drafted the American Declaration of Independence in 1776 and went on to become the third President of the US, was both a keen inventor (he came up with a revolving clothes rack, a folding ladder and a portable desk with a hinged writing board, on which he wrote the Declaration of Independence) and a big macaroni fan. This stemmed from his time during the 1780s as US Ambassador to France, when he became a big fan of European cuisine. He wrote lovingly in his diary about how macaroni was produced, including a drawing of a macaroni machine, and even had a mould for producing macaroni shipped from Naples to Philadelphia. The story about Jefferson being the inventor of macaroni cheese, is, sadly, an urban myth.

Italian-born Angelo Vitantonio patented a hand-cranked cast-iron pasta machine that both rolled and cut macaroni in New York in 1906. Three years after his death, a smaller version of the machine was patented by Angelo's son, Luigi, and it was at this point that the machine started to take off in the US. Meanwhile, in northern Italy in the 1930s, famous names in the pasta-maker business, such as Imperia and Marcato, were starting out. In the early years, founder Otello Marcato even delivered his machines by bicycle. Created in 1965, his classic Atlas 150 machine has now become an iconic Italian design. I've got one in my kitchen. I bought it a few years ago on a whim because I loved the idea of making fresh pasta, and now I use it all the time. It's a brilliantly simple design featuring two rollers – it's actually a lot like the slip rolls that I use for rolling sheet metal in my workshop. You clamp the machine to the side of the bench with a proper G-clamp, put the pasta in the top, and adjust the distance between the two rollers each time you feed the pasta through to get it thinner and thinner. All you need is 100 grams of flour and one egg per person. I mix the pasta, roll it up and let it sit while I boil the water and I've got fresh pasta in not far off the amount of time it would take me to cook the dried stuff. It tastes so much better and you've got that extra satisfaction of having made it yourself.

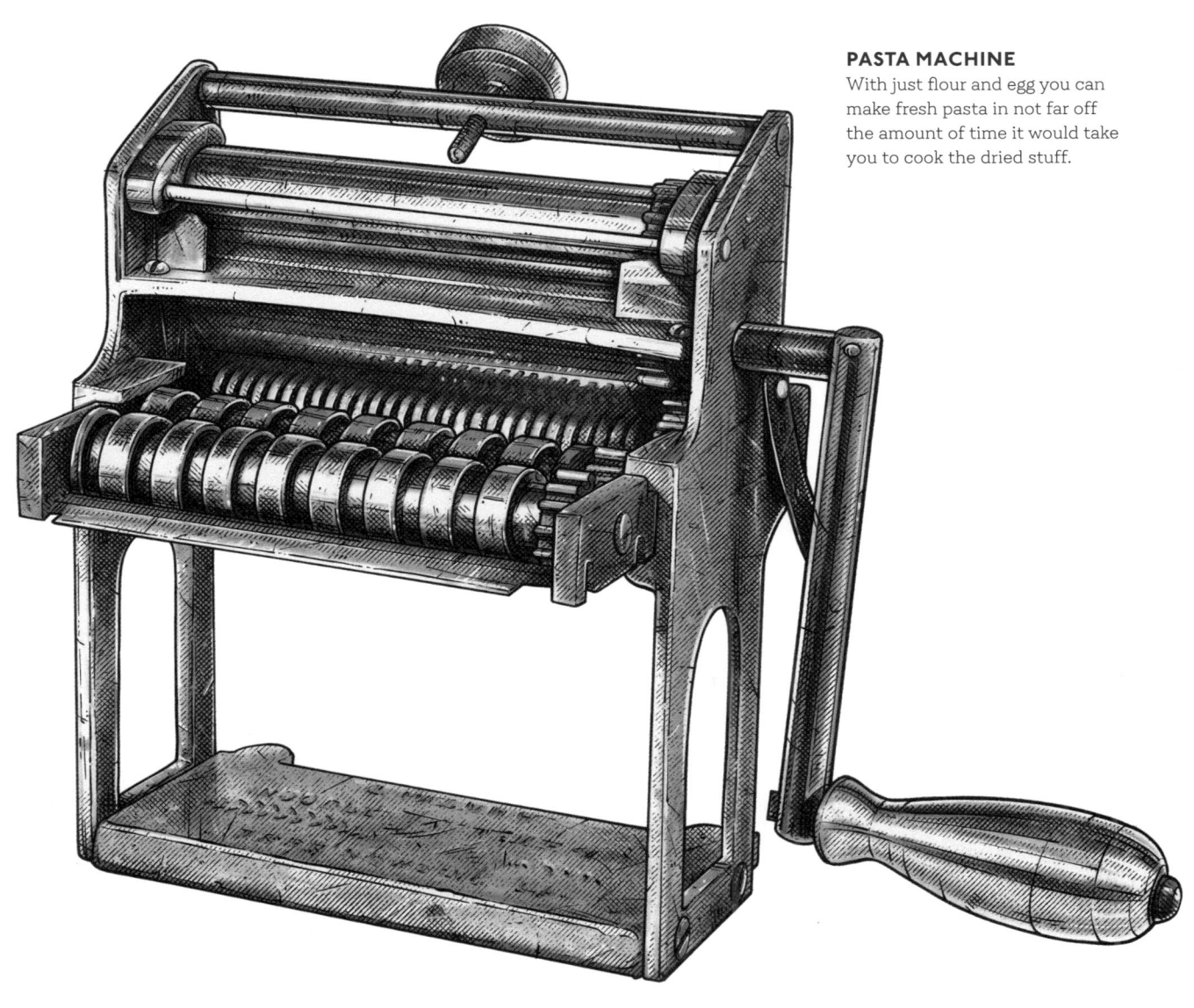

PASTA MACHINE
With just flour and egg you can make fresh pasta in not far off the amount of time it would take you to cook the dried stuff.

CREATED IN 1965, HIS CLASSIC ATLAS 150 MACHINE HAS NOW BECOME AN ICONIC ITALIAN DESIGN. I'VE GOT ONE IN MY KITCHEN. I BOUGHT IT ON A WHIM.

BEER ENGINE

MACHINE NUMBER 031

```
This is one of those machines that a lot of us are very
thankful for. So next time you're ordering a pint of ale
at the bar, raise your glass to John Lofting, the Dutch
inventor hailed by The London Gazette, in 1691, for
having "projected a very useful engine for starting of
beer, and other liquors which will draw from 20 to 30
barrels an hour which are completely fixed with brass
joints and screws at reasonable rates."
```

Lofting also developed another machine we are thankful for, in a more serious way. In the aftermath of the Great Fire of London in 1666, which destroyed over 13,000 homes, 87 churches and Old St Paul's Cathedral among other vital buildings, there was a major push to develop firefighting equipment. Lofting came up with a "sucking worm engine" in 1690, which used long leather pipes strengthened with steel wire to carry water over long distances and the power of suction to launch the water up to 120m (400ft) in the air.

The beer engine draws beer from a cask-ale keg (usually located in the cellar below the bar) along a tube and up through the spout via a long handle that you move back and forth. It does this via a cylinder with a piston inside, which is activated when you move the handle, drawing the beer from the cask via a connecting tube and out through the spout, which is often curved, hence the common name for it: swan neck. Check valves make sure the beer flows only one way. Traditionally, all the components of a beer engine were made in brass, but since 1990, the parts of the engine that actually come into contact with the beer have to be made from stainless steel or plastic. It's actually surprisingly hard to pull the handle, so it's no surprise that bartenders often alternate which arm they use.

PEPPER GRINDER

MACHINE NUMBER 032

Peugeot is said to have patented the pepper mill in 1842, although the first model they produced does not seem to have surfaced until 1874. That was their famous "Z model" pepper grinder, which became an icon.

It was originally made of plain white china, but that was replaced with a silver-plated, bakelite or wooden body, and sometimes featured a crank handle. Peugeot's pepper grinder mechanism was of very high quality, and it was used by a number of manufacturers, who made the bodies of the mills themselves.

But the history of pepper mills goes much further back than the 19th century – we're just not sure exactly how far back and at what point pepper shakers, which contained pre-ground pepper, were replaced with a mill to actually grind the peppercorns. We do know that a pepper grinder was found in the wreck of the *Mary Rose*, the flagship of Henry VIII which famously sank during the Battle of the Solent in 1545. It was pretty much intact except for the iron plates that had rusted away and it still contained peppercorns. At that time, pepper had to be imported from India, so it was a seriously luxury product in Europe.

Few pepper grinders from the 19th century survive today. In the early 20th century, they were made of a number of different materials, including crystal, sterling silver, ebony and ivory. Inside, they featured similar grinders made from steel.

The Atlas pepper mill is a particularly beautiful example of the machine, with its all-metal body (usually solid cast brass but copper, chrome and nickel versions are also produced), distinctive hand crank, fluted base, and internal hand-cut steel burrs which grind rather than crush peppercorns. The legend goes that the Atlas pepper mill is adapted from the portable coffee mills used by Greek soldiers during the First World War. I'm not entirely sure if that's true, but it's a stunning design either way. I've got both the salt and pepper versions in my kitchen. The crank handle is so much more pleasing than just twisting something. You somehow feel like you're really involved in the grinding process. The whole design is a perfect marriage of form and function.

SODA SIPHON

Bottled carbonated water was commercialised in 1773 by a name you'll be familiar with: Johann Jacob Schweppe, a German-Swiss watchmaker and keen scientist who went on to found the Schweppes company. Schweppe had built on the foundations made by legendary English chemist Joseph Priestly in the late 1760s, who'd discovered a method to carbonate water while living next door to a brewery in Leeds.

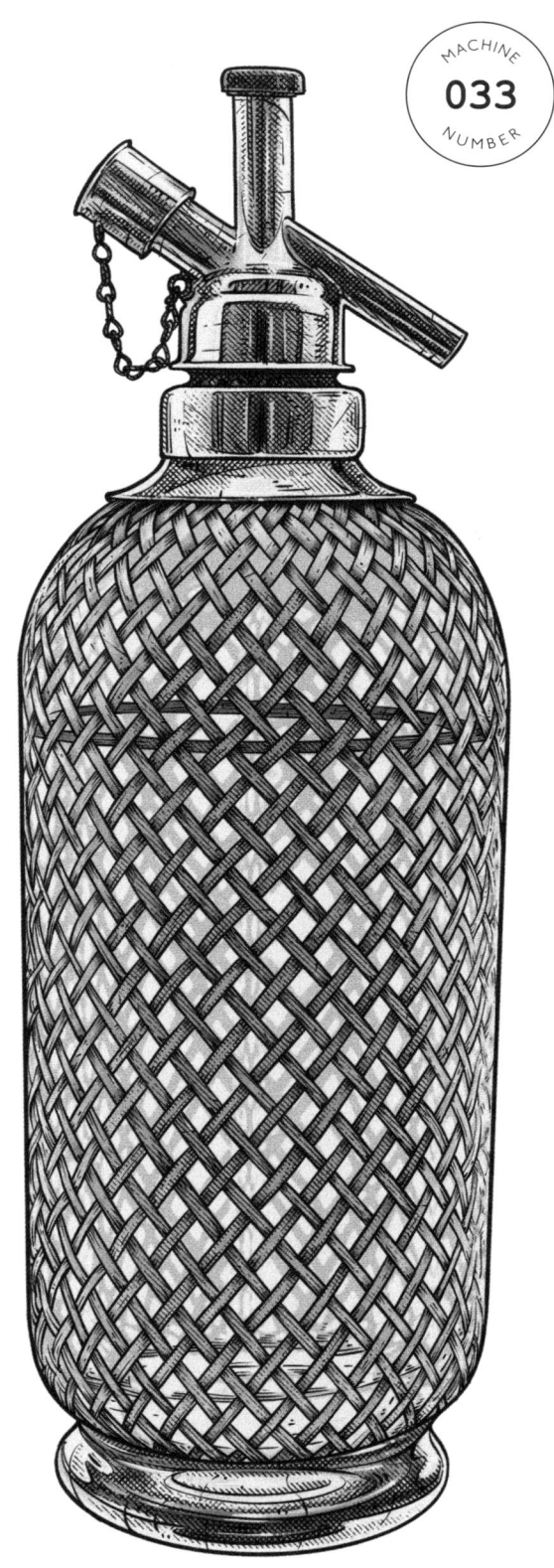

It was in 1829 that two Parisian jewellers, Deleuze and Dutillet, surprised all of France by devising a way to preserve opened champagne rather than finish it. Their invention was a hollow corkscrew connecting to an elaborate up-curved spout in the shape of a fish that could be opened and closed by a lever which controlled a valve. This allowed you to release the contents without changing the pressure inside the bottle, so the contents wouldn't go flat.

Eight years later, the precursor to the modern soda siphon bottle was invented, also in Paris, by Antoine Perpigna. It had a head mechanism that fixed onto the top of the bottle and contained a spring-operated valve and a tube travelling down into the bottom of the bottle. Opening the valve released the contents.

SODA SIPHON
Soda siphon bottles had a burst of popularity during the Prohibition era in the US. Soda helped to create drinks that masked the flavour of alcohol.

Soda siphon bottles became hugely popular during the Roaring Twenties, especially in cocktail-making in speakeasies during Prohibition in the US, which lasted from 1920 to 1933. They were both very useful at creating a drink that masked the flavour of alcohol, and at putting out fires, as a soda siphon doubles up as a very handy fire extinguisher. They are beautifully designed, reassuringly solid objects and I've had quite a few of them at home, the green and blue glass ones especially.

If you've ever wondered why they're also called seltzer bottles, the word comes from the tiny German village of Niederselters, which contains a naturally carbonated spring, formed from geothermal activity below the earth's surface, which releases carbon dioxide into the water. They bottled the water and sold it, and it became popular among the Jewish community. After Jews emigrated to the US in the 19th century, they tried to recreate the carbonated water using the soda-siphon technology and gave the drink the Yiddish name *seltzer*.

Established over 60 years.

GERAUT & CO.'S

Prize Medals awarded at all Exhibitions.

BRITISH MADE SELTZOGENES.
PATENT CLIP SELTZOGENES.

Pat. No. 11305. Metals of Pure English Refined Tin. **This patent can be fitted to all Seltzogenes.** All make Seltzogenes repaired.

CHARGES FOR SELTZOGENES.

3-pint, 5-pint, and 8-pint sizes.
We strongly recommend our charges to be used with our Seltzogenes, as they are specially prepared by us and the quality is guaranteed. 12 charges in each box.

RADIO-ACTIVE ÆRATED WATERS.

Prescribed in the treatment of Gout, Rheumatism and Nervous Affections. **RAREST WATERS KNOWN.**

PROFESSOR KLEMPERER says :—" We know of nothing better than Radium Emanation for the treatment of the worst cases of Rheumatism."

Particulars on application to—

139/141, FARRINGDON ROAD, LONDON, E.C.

"DYSPEPSALIN."

The famous remedy for many kinds of Indigestion, etc. Testimonials of cures and relief. Sample 2/6 size, post free, 2/-.

H. HAY & CO., 141, Farringdon Road, London, E.C.

NUTCRACKER

Machine Number 034

Nutcrackers might win the award for the most elaborately decorated machines in this book. Some of the most remarkable are priceless works of art, dating back thousands of years. They really put that boring-looking one in your drawer, which you only get out on Christmas Day, into perspective.

Originally, nuts would have been cracked by specially shaped stones. The first metal nutcrackers used an iron lever, and the oldest one in existence is an extraordinary looking thing dating back to the late 4th or early 3rd century BCE. It is a pair of women's forearms in gilded bronze, complete with ornate snake bracelets around both wrists. The hands are interlocked and are forced together to crush nuts by a lever operated by a handle at the base of the forearms. This nutcracker was found in 1930, near Taranto, Italy, and is now displayed in the National Archaeological Museum based there.

Iron-lever nutcrackers are the type we generally use today, although they are so boring-looking now compared to the hand-wrought elaborately decorated versions made in the past. Some of the most incredible feature dragons' jaws, ships' wheels, keys and wishing wells.

The second type of nutcracker is a screw nutcracker. The first of these would have been small, simple wooden contraptions, with screw threads carved by hand. They're usually a few inches tall – or long – and hold the nut in place while you turn the screw to break the nut open. It's actually a really clever design because unlike other types of nutcracker, it helps preserve the kernel inside. The oldest one was found in England, is made of boxwood, dates to 1631 and is a beautiful thing, featuring intricately carved decorative foliage.

The third type of nutcracker is one where you basically bash the nut with something resembling a hammer. Again, though, the decoration on some of

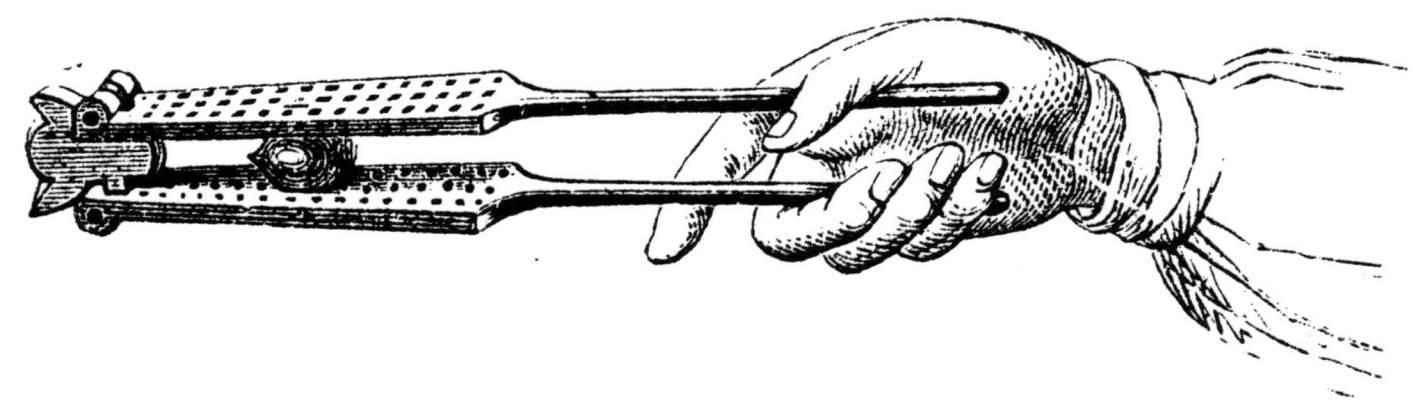

these are just incredible, looking more like small totem poles or ornate chess pieces.

As for the wooden soldier nutcracker, these were first crafted in the late 19th century. The first ones featured a handle on the figure's back, which you pushed down to move the soldier's jaw. You placed the nut in the jaws, turned the lever and you'd have yourself a cracked nut. The soldier figure was just one of the early designs, along with a policeman and a forester. It's likely they were originally conceived to poke fun at people in positions of authority.

One of my favourites is a spring-jointed nickel-plated nutcracker patented by the amazingly named Henry Quakenbush in the early 20th century. It's simple by the standards of some nutcrackers but beautiful and very well made. Look for the initials HMQ on the fulcrum. I also really like the nutcrackers that look like a G-clamp on a stand, with the T-shaped handle on top and a tapered base at the bottom. It's a simple, classic design, because, when it comes down to it, all you need a nutcracker to be is a mini-vice.

There's a serious nutcracker-collecting community out there, no more so than in Washington state, US, where you'll find the Leavenworth Nutcracker Museum, which has an absolutely unbelievable large and varied collection.

IRON LEVER NUTCRACKERS
After stones, these were the first type of nutcrackers used by humans. Originally they would have featured elaborate designs but they are far more boring now.

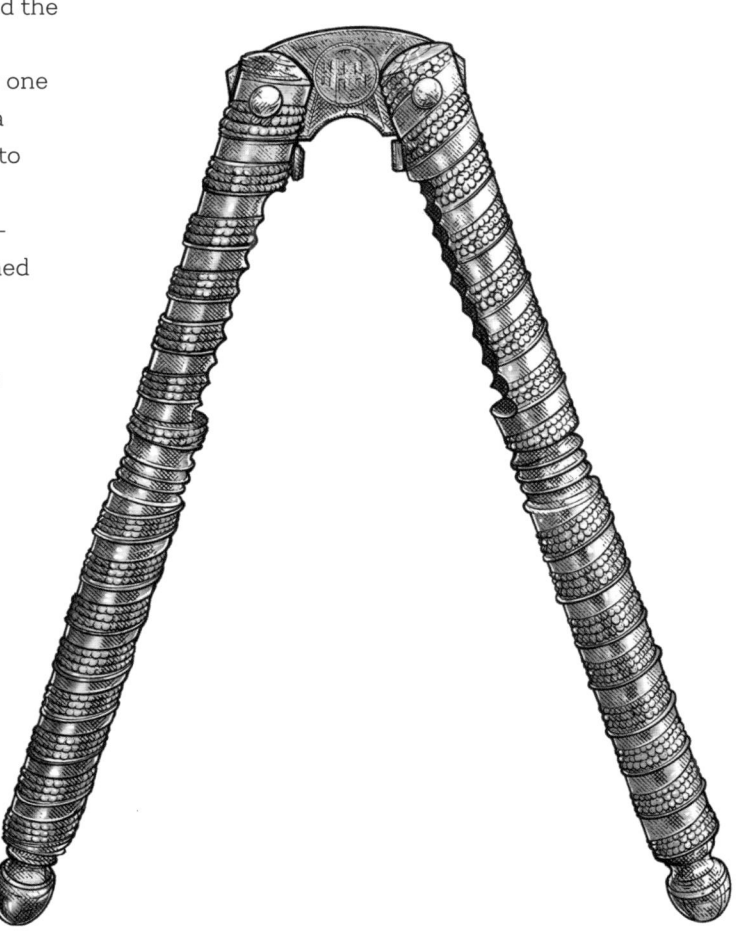

ONE OF MY FAVOURITES IS A SPRING-JOINTED NICKEL-PLATED NUTCRACKER PATENTED BY THE AMAZINGLY NAMED HENRY QUAKENBUSH.

MECHANICAL SCALES

MACHINE 035 NUMBER

The balance scales, used to compare the weights of two objects, date back to Ancient Egypt, and to at least 2600 BCE. Although no actual scales have been discovered from this time, paintings of them exist and balance weights have been unearthed.

They are a simple device, when it comes down to it, with a horizontal beam suspending two pans hung from the ends by cords. You place a standard weight in one pan and the object to be weighed in the other and adjust them until they're even.

The Romans added a pin through the middle of the beam to work as a central bearing, but other than that, little about their design changed until Leonardo da Vinci came up with a self-indicating scale in the 15th century. This was a big development because it didn't require standard weights to determine the weight of another object – the scales themselves told you the weight. French inventor Gilles Roberval came up with the next milestone with his "Roberval balance" in 1699. It featured three parallel vertical columns joined by two horizontal beams, which attached together via six pivot points. Two plates were fixed to the outermost vertical beams on the left and right. The genius of this device was that it didn't matter where on the plate you positioned the weight. On previous scales, you had to make sure the weight was in the centre of the plate or pan. His invention is still used on modern balance scales.

Things hotted up, as they tended to do, during the Industrial Revolution. In 1770, English spring-maker Richard Salter invented the spring scales in Bilston, Wolverhampton, UK. They were an inspired addition to the steelyard, a type of weighing scales involving sliding weights along a rod to balance loads suspended on chains. The spring scales work by registering the distance a spring stretches under a load. At first, they were very simple devices, but later measuring dials were added. Other than that, the spring scales are remarkably unchanged from their 18th-century origins. Salter also made a major contribution to the steam engine, developing a safety valve based on his spring principle which performed the small job of making sure the engine didn't blow up. It was essentially a spring and a ball in a tube. When the air pressure reached a certain level, the spring collapsed and opened the valve to release the pressure. Safety valves still exist today in the same form in all kinds of machines, especially those that use compressed air. Salter deserves more credit for the number of lives his invention has saved.

Salter's nephew, George, took over the company, renamed it George Salters and established a factory in nearby West Bromwich. That company (now just known as Salters), which traces back to Richard in 1760, is still a famous name in bathroom and kitchen scales in the UK. And George Salter's legacy lives on beyond scales, because, in 1878, a group of young workers at his George Salters Spring Factory formed West Bromwich Albion Football Club. Although, I'm informed that their fans famous "boing-boing" celebration doesn't have anything to do with the spring connection, which is a shame.

SPRING SCALES
Richard Salter's spring scales were invented in the UK and work by measuring the distance a spring stretches under a load.

CAN OPENER

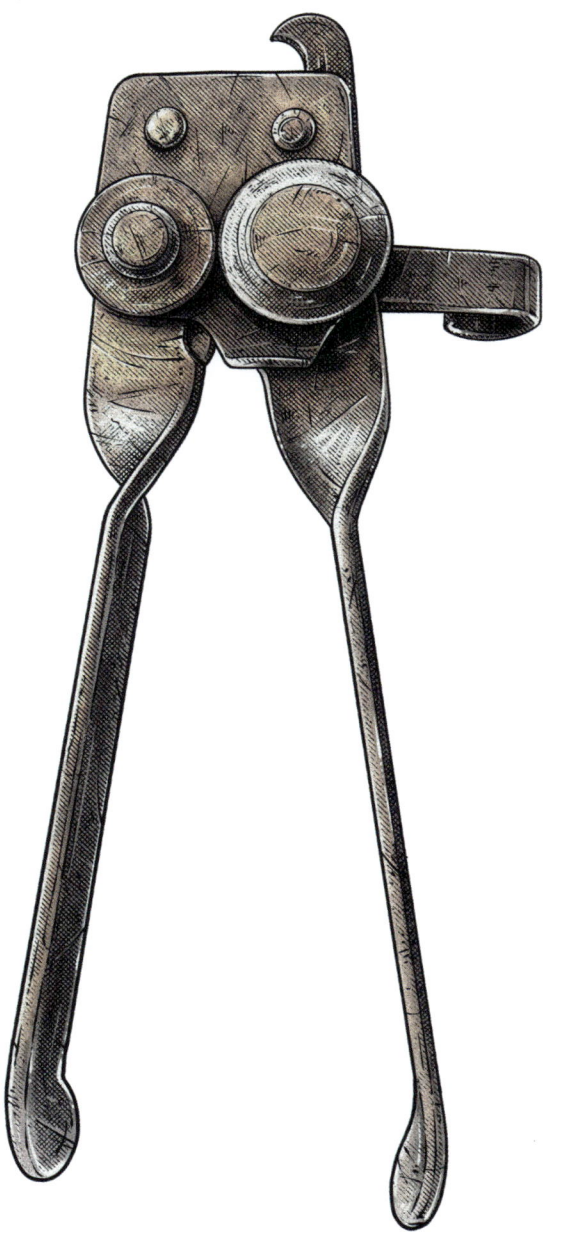

MACHINE NUMBER 036

Everyone has one of these at home. And I'd forgive you for thinking they don't seem that exciting. I'd also forgive you for thinking that both the can and can opener were invented at the same time. But their inventions were separated by nearly 50 years.

The legend of how the tin can came about involves Napoleon Bonaparte. Napoleon was looking for a method of preserving food to supply his Army and Navy during their many conflicts with other European states. So, in 1795, his government put up a prize of 12,000 francs to the inventor who came up with the solution. Nicholas Appert, a French chef and confectioner, rose to the challenge – although it would take him 14 years of experimentation to perfect his technique. And this technique was to place food in glass containers that were reinforced with wire, and sealed tightly with wax, which he then plunged into boiling water. This method worked, although the one minor issue was that no one was entirely sure why it worked. It was a full 50 years until the legendary French chemist Louis Pasteur was able to explain that the heat killed the bacteria in the food and the strong seal prevented other microorganisms from sneaking into the jars. So Appert claimed his prize in 1809 or 1810 and used the cash to invest in a preservation technique that had just been pioneered in England by the inventor Peter Durand: the tin-plated iron can. (Although, having said that, a canning operation does seem to

CAN OPENER
The can opener as we know it today was invented in the 1920s, more than 100 years after the first cans were used for storing food.

have been going on in the Netherlands since the middle of the 18th century, with the Dutch Navy preserving salmon in tin canisters.) Durand's method involved rolling sheets of tin-coated wrought iron into a cylindrical shape, and then soldering the top and bottom onto the body. The trouble was that they were about as thick as a fence, so weren't exactly easy to access unless you had a hammer and chisel to hand.

By the 1850s, the iron can was being replaced with thinner, steel versions. And that's when the can opener came about, courtesy of Ezra J. Warner of Waterbury, Connecticut. Warner developed a device with a blade that could pierce through the lid of a can. A guard stopped the blade from piercing too far downwards, and then you used a second, sickle-shaped blade to cut around the top of the can, using a sawing action, and hey presto – you were in. The slight pitfall was that after you'd got in, your reward might be a severed finger courtesy of the lethal jagged metal edges that the contraption left in its wake. Warner's device was used a lot during the American Civil War (1860–1865), but it wasn't really a hit at home.

Attempts were made to develop a device with a rotating cutting wheel around 1870, but it wasn't until the 1920s that American inventor Charles Arthur Bunker developed the genius, tooth-wheeled design, with the pliers-like handles, where you hold it with one hand and turn a key with the other. Bunker was awarded a patent in 1931 for his can-opening device and it was produced by the Bunker Clancey Company of Kansas City, that same year. And that model has basically exactly the same mechanism as the one sitting in your kitchen drawer right now.

CHAPTER

3

HERITAGE CRAFT

MACHINE NUMBER	MACHINE NAME	PAGE NUMBER
037	LATHE	078
038	TYPEWRITER	080
039	PAPER CUTTER	084
040	SEWING MACHINE	086
041	HYDRAULIC PRESS	092
042	RIVET-MAKING MACHINE	094
043	RIVETING MACHINE	098
044	LETTERPRESS	100
045	BRILLIANT-CUT MACHINE	102
046	WHEELING MACHINE	104
047	SCAIF	109
048	PHARMACIST'S PILL ROLLER	110
049	CIRCLE-CUTTING MACHINE	112
050	ENGINE CRANE	114
051	CORK PRESS	116
052	BOOKBINDING MACHINE	118

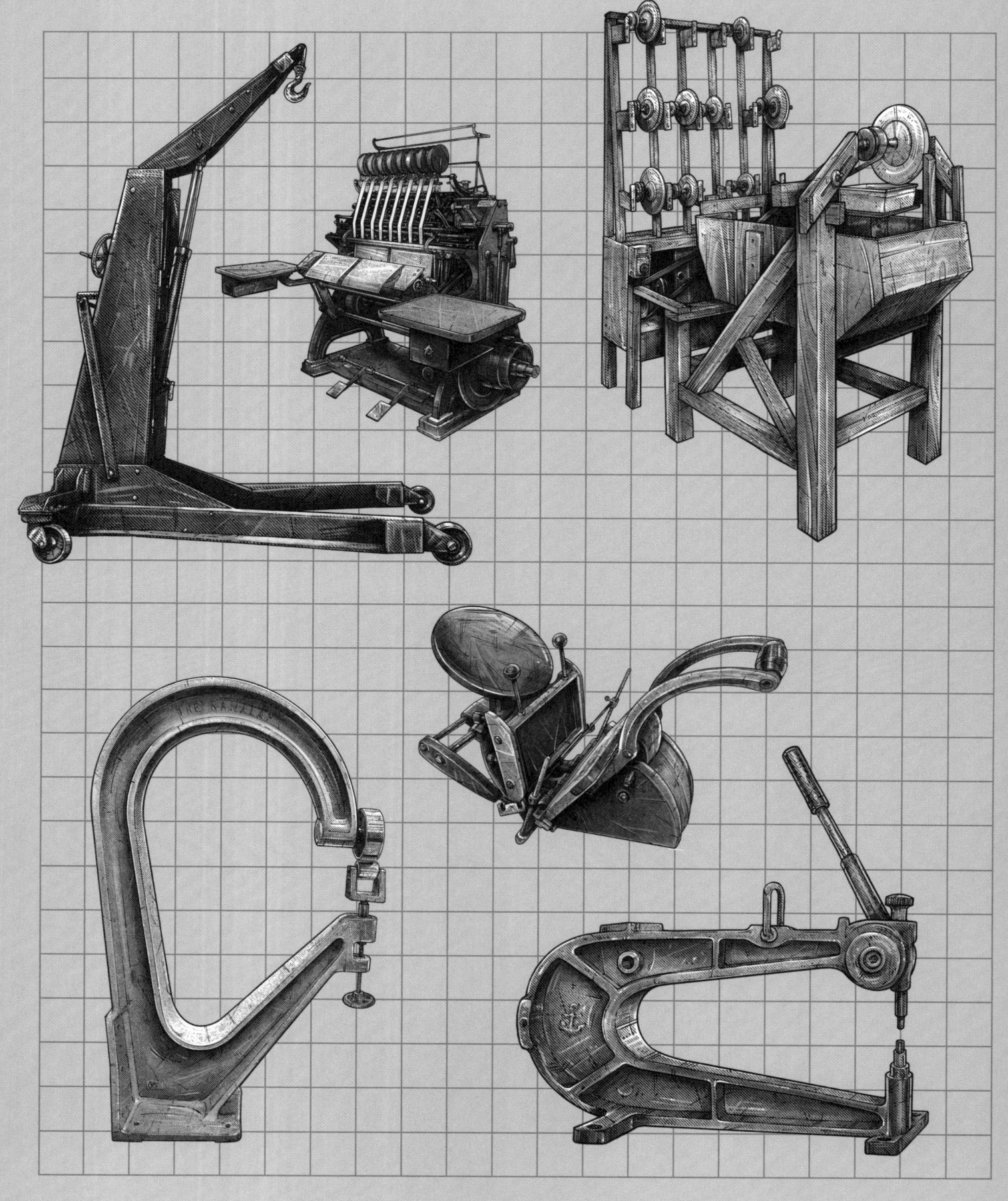

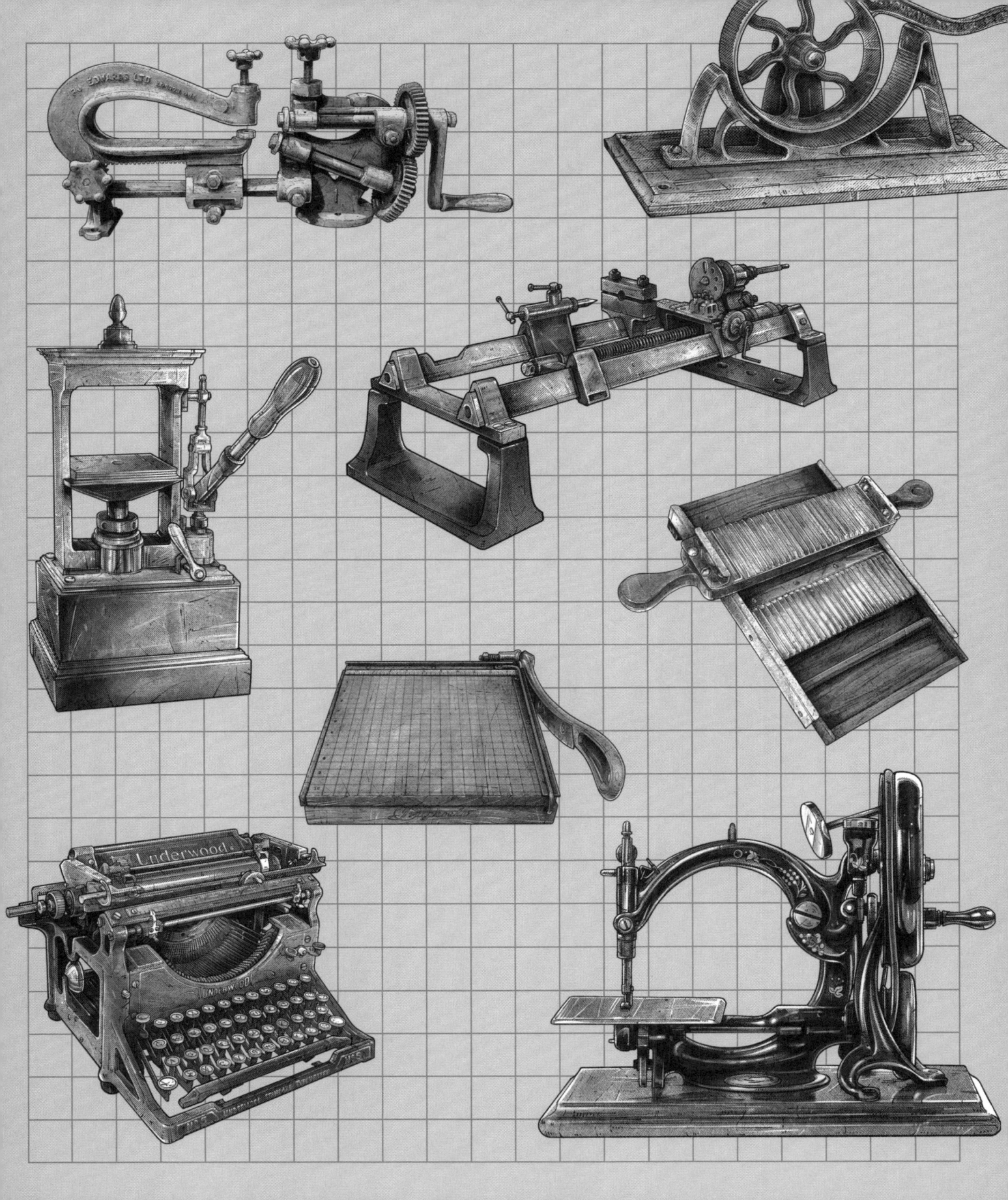

LATHE

The lathe is a machine tool (a mechanical device to cut and shape metal, wood and other materials) that rotates a workpiece so you can shape it with a fixed cutting or grinding tool. It's a surprisingly ancient invention: there's evidence for a woodworking lathe operated by two people going back to Ancient Egypt.

It wasn't very sophisticated, though, with one person turning the wooden workpiece using a rope and the other shaping it, but it was better (and safer) than manually turning it with your hand. The first lathes would have been ground-level devices, with two parallel wooden supports from which iron spikes projected to hold the workpiece in place. A strap wrapped around the workpiece gave the turning power, and this would have been controlled by an assistant because it needed two hands to operate.

The next design was the bow lathe. String attached to the ends of a bow (think bow and arrow) wrapped around the workpiece, which meant a sole user could turn a workpiece with one hand by moving the bow

LATHE
Lathes have existed since Ancient Egypt as a tool to shape wood or metal. Leonardo da Vinci had a hand in its development in the 15th century.

backwards and forwards. This meant their other hand was freed up to use whichever tool was needed. The sacrifice was the amount of power this could generate, though, using just one hand for each task, but bow lathes were still in use right up until the 1800s.

The next milestone came along with the pole lathe, sometime during the Middle Ages, which consisted of a framework that lifted the bed of the lathe up, and a long, thin, flexible pole extending out above the lathe. A string extended from the tip of the pole that wrapped around the workpiece and down underneath the lathe where it was fastened to a foot pedal (treadle). This was a big development, as it left two hands free again.

Leonardo da Vinci played a part in the history of the lathe, making a sketch of a lathe operated via a crank, treadle and flywheel around 1480. It represented a big leap because it supplied continuous motion rather than reciprocal motion. The British craftsman and woodturner Stuart King was commissioned to bring Leonardo's design to life for an exhibition in 2012, and he proved it not only worked, but worked very well.

The French inventor Jacques de Vaucanson, who'd made his name constructing life-size mechanical creatures (and people), developed the metal slide-rest lathe (a slide rest being an attachment that held the cutting tool firmly in place) sometime between 1750 and 1760. His lathe was designed specifically to create the shimmering effect you achieve with silk when you crush it, which he did by passing the fabric evenly through two copper cylinders.

And that brings me on to the English engineer and inventor Henry Maudslay, who did not invent the metal lathe (or the slide rest), as the internet seems to believe, but he did produce a screw-cutting lathe around 1798. He wasn't the first to make such a machine, but his was the best-known and became the most influential. In fact, his original screw-cutting lathe is such a big deal that it's in the Science Museum in London. The reason why it was such an important invention is because it meant you didn't have to custom-build screws and bolts. They became interchangeable parts, and that opened up the door to mass production.

So for a very brief, blissful time after Maudslay's invention around the turn of the 19th century, there would have been only one standardised thread size. Imagine! Of course, after that, hundreds of different thread sizes were made by lots of different manufacturers. This was followed by attempts to unify the thread system, but that actually caused even more thread sizes to be made.

Whenever something comes in to *The Repair Shop*, especially the handmade stuff, there's that anxious moment when I take the object apart and find that a screw, nut or bolt is missing (or that a screw thread has worn away, or the hole's become too large). At that point, I become a detective. First, I take out my thread-pitch measuring gauge to work out how many threads per inch there are. After that, I start opening page after page of old charts to try and find out the manufacturer. At that point, I might get lucky because I might have already bought the random screw tap (which you need to create internal threads in a hole or external threads on a screw) for a previous *Repair Shop* job. If not, I end up on the phone to an obscure supplier that makes a very specific screw tap. The amount of work that goes into finding or making a single screw is astonishing sometimes. But as frustrating a process as it can be, there comes that moment where you have something that fits. It's the metalworker equivalent of the glass slipper finally fitting Cinderella. Yes, it would be a lot easier to drill the whole out and use a modern screw, but I'm in the business of preserving the originality of a unique item. Otherwise you lose a bit of magic.

I actually have a big metal lathe in my workshop which has the ability to cut threads but I've never been brave enough to use it properly. Screw-cutting and thread-cutting is such an art and I'll get there eventually, when I have a half-day free and I sit down and say, "Right – the time has come. I'm making a bolt today."

TYPEWRITER

MACHINE NUMBER 038

In 1575, an Italian typographer and inventor by the name of Francesco Rampazetto is said to have invented a machine that could impress letters onto paper. It was a basic machine made with cubes of wood, and was designed to help blind people communicate. This seems to be the precursor to the typewriter, but there isn't any actual evidence the machine existed except for a reference to it in a magazine from 1924.

The first person to receive a patent for a mechanical writing machine was Englishman Henry Mill, in 1714. The application is worth mentioning, not least for the amusing language used by Mill's representative: "Whereas our trusty and beloved subject, Henry Mill, hath, by his humble petition, represented unto us, that he has, by his great study, pains and expense, lately invented and brought to perfection an artificial machine or method for the impressing or transcribing of letters singly or progressively one after another, as in writing, whereby all writings whatsoever may be engrossed in paper or parchment so neat and exact as not to be distinguished from print. That the said machine or method may be of great use in settlements and public records, the impression being deeper and more lasting than any other writing, and not to be erased or counterfeited without manifest discovery."

Wow! At that point in time, though, you didn't have to actually submit any drawings along with your fancy language, so we don't know what it looked like, and it's likely the machine wasn't ever produced.

We have to fast-forward to 1867 for the next major development, as this was the year that the first practical typewriter was developed by Christopher Latham Sholes, an editor, publisher, politician and inventor from Pennsylvania, US, although he settled in Milwaukee, Wisconsin. Sholes worked with his friend, and printer, Samuel W. Soulé on the design for a page-numbering machine, which they were granted a patent for in 1864. But it was lawyer turned inventor and fellow Milwaukee resident Carlos Glidden who suggested to Sholes that his page-numbering machine might be adapted into a letter-printing machine. Glidden pointed Sholes to an article in the July 1867 issue of the journal *Scientific American* about the "pterotype". This was a prototype writing machine developed by John Pratt, a lawyer and editor from Alabama, who had spent the latter years of the 1850s getting infuriated with the bruises on his fingers caused by spending so much time writing.

However, as with the history of the sewing machine, war would ring the changes – for Pratt, at least. The American Civil War (1860–1865) meant that Pratt struggled to find anyone to invest in his design, so he sold up, and moved to London, England, in 1861, to try his luck there. He completed his invention in 1864, presenting it to three learned institutions in London: the Royal Society, the Society of Engineers and the Society of Arts. He was granted UK and French patents for his invention in 1866.

It was more like a printing press than a modern typewriter but it did contain some unique ideas, including a type plate which moved vertically and horizontally by pressing the keys and a hammer that struck the paper from the rear against the keys.

TYPEWRITER
The Underwood front-striking typewriter from 1896 was a game-changer, as it allowed you to see what you were typing without having to stop.

"CLOSE YOUR EYES AS YOU TOUCH TYPE AND YOU ARE A BLACKSMITH SHAPING SENTENCES HOT OUT OF THE FORGE OF YOUR MIND." –TOM HANKS

Meanwhile, Sholes was inspired by what he'd read and got to work at once on a smaller, simpler and more user-friendly version of Pratt's idea to sell in the US. He, Soulé and Glidden were granted a patent for the second version of their machine, which had the keys laid out in alphabetical order over two rows. The name for it – the typewriter – he'd got from the description of Pratt's device in the *Scientific American*. Fittingly, Sholes and his team used the typewriter to write letters to potential investors, including James Densmore, who bought a 25% stake (stumping up the cost of the entire development of the project to that date) despite never actually having seen the machine. The inventors also sent prototypes of the machine to various professionals inviting their feedback, including renowned stenographer James O. Clephane. One of the changes Sholes made as a result of the feedback was to separate the most commonly used letters to stop the machine from jamming. This problem occurred because weights swung the typebars back to their "rest" position, so if keys were tapped in quick succession, their corresponding typebars could collide and get stuck. It wasn't enough to just reorder random keys, though. So Sholes and his team analysed how often different letters were used. The upshot was that a new key pattern was designed to separate frequently occurring alphabetical pairs, like A and B, and S and T. They also scattered the most commonly used letters. Their solution eventually led to one of the most significant breakthroughs in the history of the typewriter (and modern technology for that matter): the QWERTY keyboard layout.

Densmore had started selling the Sholes and Glidden Typewriter from 1871 and although the machines were impressive, the company was struggling to make a profit. It needed a much bigger operation to increase production and reduce its production costs per unit. Densmore was encouraged by a manufacturing acquaintance to approach the renowned firearms maker E. Remington and Sons, which he did via a typewritten letter. This worked wonders and Densmore was invited to the Remington offices in New York, in 1873, to demonstrate the typewriter. At that point, the keys in the top row (not including the row of numbers) were: Q W E . T Y I U O P. Following that meeting, Remington asked to buy the patent from Sholes and Densmore. Sholes sold his stake for $12,000, while Densmore negotiated a royalty, which eventually earned him $1.5 million. Remington acquired manufacturing rights and dedicated an entire wing of its factory in New York to building typewriters. Its mechanics made some adjustments to the key layout, including changing the full-stop key to an R, so the row now started Q W E R T Y. The story goes that this change was made so that Remington salesmen could spell out "typewriter" quickly using the keys from just the top row, but that sounds like a good old-fashioned urban myth to me.

The first machines were sold in July 1874 and were renamed the Remington. Among its early fans was the famous author Mark Twain, who proudly stated that in 1876, he was the first person to submit a typewritten novel (which he remembers as *The Adventures of Tom Sawyer*) to a publisher. Although, thanks to the research of historian Darryl Rehr, it seems to be a slightly dodgy claim, as he dictated it to an assistant he'd hired and the book was actually not *Tom Sawyer* but a memoir called *Life on the Mississippi*.

In 1878, Remington released its No. 2 model, which became the first hugely successful typewriter. The No. 2 was the first typewriter to feature a shift mechanism, allowing the typist to use both upper- and lower-case letters. As with previous typewriters, it was an upstriking machine, so the keys rose up, striking the paper through the ink ribbon. The downside was that you couldn't see the text as you typed – you had to manually raise the hinged carriage that held the paper to see what you'd written.

This problem was solved around 1892 by Franz X. Wagner, with the development of the front-striking typewriter. This featured typebars laid out in a semicircle in front of the carriage, so that when you pressed a key, it struck the front of the platen (roller), so you could see what you were typing. But Wagner wouldn't achieve commercial success with his machine for a few years yet. Meanwhile, the Daugherty "Visible" front-striking typewriter was released in 1893. It's a beautiful machine, but it didn't perform as well commercially as perhaps it should have. In 1895, Wagner sold his designs and his business – the Wagner Manufacturing Company – to businessman John T. Underwood, a producer of ribbons and carbon paper who was looking for a new opportunity after Remington had begun to produce its own ribbons. The following year, Underwood released its Model One typewriter (which had Wagner's name stencilled on the back) and it was an absolute game-changer. Although it had similar features to the Daugherty, it was much better engineered and sold in its thousands. It essentially made all upstriking typewriters produced to this point obsolete.

The Underwood Model 5, introduced in 1899, became the market leader for almost 30 years. Five million of them were sold in that timeframe – they were said to be producing one a minute from their factory in Hartford, Connecticut – and almost every other manufacturer imitated the design. Underwood was eventually acquired by Italian company Olivetti in 1959, itself a highly respected typewriter producer. Meanwhile, Thaddeus Cahill, who'd already invented the "telharmonium" – the world's first electromechanical musical instrument – in 1896, was working on an electric typewriter, which he first produced in 1899 or 1900. But it wasn't until 1933 that the first really successful model was produced in the US, by IBM, which had acquired the tools, patents and facilities of Electromatic typewriters the previous year and spent an incredible $1 million on a redesign.

And while electric typewriters ultimately took over, enthusiasm for manual typewriters has surged in the last 20 years with the appearance of the maker movement and the revival of heritage craft. Tom Hanks might be the highest-profile typewriter enthusiast in the world. His passion started when a friend gave him his first typewriter when he was 19. It was a cheap plastic one, but when he took it into a machine shop to get it serviced in 1978, the owner of the store showed him his collection of portable typewriters. Hanks was blown away and he ended up buying a vintage Hermes model for $45. It was the beginning of a love affair, and at one point in his life, Hanks had amassed 250 machines. He sees them as "a brilliant combination of art and engineering", and in the trailer for the documentary *California Typewriter* (2017), he said something which really resonated with me: "Close your eyes as you touch-type and you are a blacksmith shaping sentences hot out of the forge of your mind." Amazing. He doesn't admit to having a favourite typewriter, but he clearly has a special place in his heart for the brands Smith-Corona and Hermes, and also said: "I have a thing for my Olivetti Lettera 22s, as they are masterpieces of design, the action is crazy fast and light, and the typewriter is in the Museum of Modern Art."

SHOLES USED THE TYPEWRITER TO WRITE LETTERS TO POTENTIAL INVESTORS.

PAPER CUTTER

MACHINE NUMBER 039

The precursor to the modern paper cutter, or guillotine, was patented in 1844 by French inventor Guillaume Massiquot, but there were earlier versions dating back decades before that — only they were about the size of a large desk and used by commercial printers.

One of the most well-known companies that produced this earlier type of paper cutter (as well as all sorts of printers', bookbinders' and stationers' machinery and materials) was Harrild & Sons of Farringdon Street, London, which was established in 1809. It counted famous newspapers including *The Times* and the *Morning Herald* among its customers, as well as the government and Eyre & Spottiswoode (printers to Queen Victoria). It moved to Fleet Works on Farringdon Street in the 1850s following its successful showing at the Great Exhibition of 1851, which was held in Hyde Park, London in the specially designed Crystal Palace. The ornately designed Fleet Works building is still standing and, from the outside, is almost exactly the same as it was with its beautiful terracotta detailing. The company went out of business in the late 1940s and the building is now a bar, which the owners have named Harrild & Sons in a nod to its famous past. Harrild & Sons illustrated catalogues come up occasionally on eBay and they're a real thing of beauty.

Milton Bradley, the legendary American board-game manufacturer and business magnate, is often mistakenly credited with inventing the paper cutter. He did file a patent for an improved version of a hinge-operated, one-armed paper cutter in 1878, which was smaller, more portable, cheaper to produce and claimed to be "well adapted for cutting long strips from card-board as would a much larger machine." Bradley had established the Milton Bradley Company in 1860, which went on to make some of the most famous board games of all time, including *Battleship, Buckaroo!, Connect Four, The Game of Life* and *Twister*.

The Ideal School Supply Company, based in Chicago, produced several paper-cutter models in the early 20th century, including its Ingento No. 3 model complete with a heavy steel or cast-iron handle and lined maple cutting board. Beautiful but deadly. Everything changed around the 1960s with the invention of the rotary cutter, with a circular cutting blade encased in a metal (or plastic) head that moved

THE ROTATRIM BECAME A FEATURE OF PRETTY MUCH EVERY SCHOOL, COLLEGE AND LIBRARY IN THE UK. I HAVE A VERY SPECIFIC MEMORY OF USING ONE.

along a metal track. The most famous rotary cutter in the UK became the Rotatrim, developed by designer Alan Hall in 1966 for a photographic show. The Rotatrim became a feature of pretty much every school, college and library in the UK. I have a very specific memory of using one, which I'd restored so that I could cut my first business cards after I'd printed them on my Adana letterpress. The unique sliding and shearing sound it made was very satisfying.

PAPER CUTTER
With its heavy steel or iron handle, the Ingento paper cutter was a deadly piece of machinery. It was a safer world all round when the rotary cutter was invented.

SEWING MACHINE

MACHINE NUMBER 040

It might surprise you to learn that the origins of the mechanised sewing machine involve angry torch-wielding mobs; one man's conscience winning out over a huge potential pay day; one of the unluckiest guys on the planet, who managed to turn things around; and a savvy former travelling actor who became one of the richest people in the early 20th century.

The first British patent for a mechanised sewing machine was granted to a London-based German inventor and doctor by the name of Charles Fredrick Wiesenthal in 1755. It was a simple machine, featuring a needle with a point at both ends and an eye at one end. The design had the needle being passed through the fabric using mechanical fingers and grasped by another pair on the other side. There isn't any evidence that this machine was actually produced though, and it wasn't until 1790 that another design for a sewing machine – this one, operated via hand crank – was patented by Thomas Saint, a cabinetmaker from London. He's got nothing to do with the AllSaints fashion brand and its shop windows full of sewing machines, by the way, which is a shame.

To be fair to Thomas Saint, he actually patented two different sewing machines and five different types of varnish, describing them all in his patent application (which sounds like it's straight from a Jane Austen novel) as: "An Entire New Method of Making and Completing Shoes, Boots, Spatterdashes, Clogs, and Other Articles, by Means of Tools and Machines also Invented by Me for that Purpose, and of Certain Compositions of the Nature of Japan or Varnish, which will be very advantageous in many useful Appliances." Spatterdashes are something Mr Darcy would have worn to protect his boots and stockings from the ravages of mud, in case you're wondering. Again

though, Saint's machine probably wasn't actually made at the time, but in 1874, a sewing machine maker unearthed the design while digging through the Patent Office, built the machine and demonstrated that it did actually work (after a fair amount of tinkering). That replica is now in the Science Museum in London. Saint's design was really significant, though, because it included a few elements that are found on modern machines: an overhanging arm, a feed system for the fabric and a form of tensioning.

Now to the tale of Monsieur Thimonnier, a French tailor who was granted a patent for a new type of sewing machine in 1830. Everything was going well for Thimonnier. The French government loved his invention and gave him a contract to produce uniforms for the French army, who were busy trying to conquer Algeria. Thimonnier got himself a workshop in Paris, which housed 80 of his machines. Unfortunately for Thimonnier, he wasn't as popular with the Parisian tailors as he was with the government, and an angry mob broke into his workshop and destroyed everything in 1831. Thimonnier barely escaped his workshop with his life, let alone any of his machines. He wouldn't let that get in his way, though, and designed a new version of his machine, but the torch-wielding mob hunted him down again. This time, he managed to save one sewing machine, and eventually fled across the Channel to England, but wasn't able to achieve any further success and died penniless. Despite his sad end, Thimonnier's 1830 design did make some serious progress: his machine featured a horizontal arm mounted on a vertical bar and used a hooked needle that was drawn from a reel underneath to create a chain stitch on the top surface of the fabric. And his was the first practical and operational sewing machine that was available commercially. Plus, an example of his 1830 sewing machine is also in the Science Museum in London. Not that that will come as any comfort to the poor guy, though.

The next major development happened across the Atlantic in 1832–1833 courtesy of a remarkable mechanical engineer called Walter Hunt. This guy was a genius, and invented the safety pin among countless other devices. He came up with a sewing machine featuring two needles, one with an eye in its point, and two threads, which interlocked, creating the first "lockstitch" (now the most common stitch created by a sewing machine and named for the entwining shape the two threads form). But Walter Hunt wasn't happy with the knowledge that his invention would put tailors and seamstresses out of a job, so he never filed for a patent.

In 1844, the English inventor John Fisher designed what might be regarded as the first modern sewing machine, combining a number of needles and shuttles that worked simultaneously to lock two threads together and produce a lasting stitch, although it was designed for ornamenting lace. However, quite unbelievably, Fisher's patent was misplaced by the Patent Office and so he was completely overlooked in the legal battles that followed. The poor chap was never recognised for his achievement.

Meanwhile, 3,000 miles away, in Boston, Massachusetts, Elias Howe Jr, an apprentice to a maker of mariner's tools and scientific instruments, began working on a similar idea to Hunt and Fisher. Howe Jr was awarded a patent in 1846 and even staged a man versus machine contest to test his invention and attract publicity. The machine beat five seamstresses but, amazingly, the display still wasn't enough to register even one sale. Having poured all his money into his invention, Howe Jr was now in serious debt. He thought he might have better luck across the pond and asked his brother Amasa Howe to travel to London with the sewing machine to attract an investor. Amasa succeeded in piquing the interest of a corset maker called William Thomas, and Howe Jr travelled over to join him. The arrangement didn't work out, though, and the two brothers sailed back to America, selling everything they had to pay the fare. Now penniless, Howe Jr arrived home to find his wife dying. And the bad news kept coming because while he'd been in London, the mechanised sewing machine had massively taken off in the US, and what's more, they were essentially rip-offs of his own design. Somehow, at absolute rock bottom, Howe Jr found the strength to fight all of the manufacturers who had infringed his copyright and he won every one of the ensuing court cases. Then, up stepped a man with a plan to join forces with Howe Jr. And that man has a surname everyone will be familiar with: Singer.

THE MACHINE BEAT FIVE SEAMSTRESSES BUT, AMAZINGLY, THE DISPLAY STILL WASN'T ENOUGH TO REGISTER EVEN ONE SALE.

THE WILCOX & GIBBS SEWING MACHINE
Early Wilcox & Gibbs machines were made of cast iron and shaped like the letter "G" for Gibbs.

THE WHOLE INDUSTRY BECAME EMBROILED IN A LENGTHY PATENT LITIGATION BATTLE, WHICH WAS LATER KNOWN AS THE SEWING MACHINE WAR.

Isaac Singer, a young travelling actor from New York, found work as a machinist to pay the bills. He discovered he was very good at it, and even better at coming up with ideas for new machines. In 1839, when he was in his late 20s, he was awarded a patent for a rock-drilling machine which he came up with while helping to dig a waterway in Illinois. Ten years later, he had moved to Boston and was asked to come up with an improved version of Howe Jr's sewing machine design. Among his innovations, he came up with the foot-treadle to continuously feed the fabric and a needle that moved up and down rather than side to side. I. M. Singer & Company was formed in 1851, co-founded by a very savvy lawyer called Edward Clark, who was behind its advertising campaigns.

But surrounded by multiple inventors and companies all securing patents for various sewing machine parts, the whole industry became embroiled in a lengthy patent litigation battle, which was later known as the Sewing Machine War. Singer was sued by Howe Jr in 1854 for using his lockstitch and ended up having to pay Howe Jr a lump sum for the machines Singer had already produced which infringed Howe's patent. This didn't stop Singer's machine being awarded first prize at the Paris World's Fair the following year, though. The world was taking notice of Singer's machine and his company.

Meanwhile, back in the States, all the manufacturers were losing a lot of money fighting court battles. So one of them, Orlando B. Potter, came up with an idea to put an end to the war. He proposed that several of them pool their patents (basically an agreement between patent owners that they could all use each other's parts). This plan worked out – becoming the first patent pool in history – although you might think a bunch of inventors could have come up with a more inventive name for the association than the "Sewing Machine Combination", but you can't have everything. And they literally didn't have everything because one vital component of everyone's sewing machine was Howe Jr's patented lockstitch and eye-pointed needle. So the Sewing Machine Combination went to Howe Jr with cap in hand to cut a deal rather than pay him a royalty fee of $25 for each machine they made that infringed his patent. They stitched together a deal and Howe Jr eventually received $5 for every machine they sold, which didn't end up being a bad deal bearing in mind that 110,000 sewing machines were sold in 1860 in the US alone. Howe Jr died in 1867 having earned close to $2 million (over $40 million in today's money). What a turnaround.

As for Isaac Singer, in 1867 he'd just opened a factory in Clydebank, just west of Glasgow, Scotland, and the recently renamed Singer Manufacturing Company became the first overseas company in the world. Calling the Clydebank site a factory downplays it a little, though. It was more like a city, sprawled over a 46-acre site, the crown jewel of which was a huge clock tower emblazoned with "SINGER" above the clockface. Railways were built to link the different departments and a station was later constructed to help transport workers from Glasgow reach the factory. In fact, Singer station is still there today serving commuters to and from Glasgow.

The Clydebank Singer factory was some operation, and its construction was overseen by renowned building contractor Robert McAlpine, the Scottish businessman whose company went on to undertake a number of landmark buildings including Wembley Stadium, the Glenfinnan Viaduct (the one in *Harry Potter*), the National Theatre, the Eden Project and London's Olympic Stadium. The factory even had a water-sprinkler system, making it one of the most modern in the world.

By 1913, the Singer company was making 3 million sewing machines a year, although everything was about to change. The outbreak of the First World War the following year meant that the Clydebank Singer Factory was converted to make munitions, including 300 million artillery shells and more than 350,000 horseshoes. After the war ended, it cranked up production of its sewing machines and in 1921 produced a model powered by an electric motor. But this innovation wouldn't reward sewing-machine makers, because everything had changed after the war. Women had the vote in the US and UK and the role of the traditional homemaker was evolving.

The advent of the Second World War stopped sewing machine production in favour of weaponry and other wartime products. In 1945, Singer went back to sewing machine production, but its cast-iron models were falling out of favour. The Clydebank factory stopped producing the classic cast-iron Singer sewing machines in the 1960s in favour of aluminium versions. The end of the 1960s saw a downturn in Singer's fortunes and unsuccessful diversification projects and a fall in the number of sewing-machine sales led to the Clydebank factory closing in 1980. It was demolished in 1998 and the company went bankrupt the next year.

Singer completely dominated the sewing machine market for generations. Their model with the cast-iron stand and foot treadle, first produced in 1856 and known then as the Turtle-Back, has rightly become a manufacturing icon. But it's not the make that I've got in my workshop. There were other incredible manufacturers out there in the 19th and 20th centuries. And that brings me on to Wilcox & Gibbs, a make I discovered when a wreck of a sewing machine arrived on my workbench at *The Repair Shop*.

The American farmer and inventor James Gibbs patented a chainstitch single-thread sewing machine with a rotating hook in 1857. He joined forces with investor James Wilcox, after a chance meeting, to form Wilcox & Gibbs Sewing Machine Company. Wilcox was a clever businessman and hired the renowned Rhode Island-based clockmaker Brown & Sharpe to manufacture the machines for them. This was a stroke of genius, as Joseph Brown wasn't just the co-owner of a successful firm – he was also an accomplished inventor who went on to develop several unique machine-tools to make Wilcox & Gibbs sewing machines. Although the first 50 sewing machines took several months to make, the hard work paid off as Wilcox & Gibbs established a reputation for reliability, precision engineering and near-silent operation. They also hired a chap called John Emory Powers – the world's first full-time copywriter and the man widely regarded as the father of modern creative advertising. He came up with a plan for buyers to try the machines, a payment instalment plan (as the machines weren't cheap) and pioneering marketing ideas such as taking out full-page ads cleverly written as short stories. Wilcox, Gibbs, Brown, Sharpe and Powers were an absolute dream team – five guys at the top of their game. And so perhaps it's no surprise that the orders came flying in. So much so that the company couldn't actually keep up with the demand.

With so many orders, Wilcox & Gibbs didn't hang about. It saw the potential for sales in Europe and had offices in Regent Street in central London up and running by 1859. However, as with Singer, it was about to be derailed by conflict, but this time it was the American Civil War (1860–1865). And this posed a serious problem to the two partners, as they suddenly found themselves on opposing sides. Gibbs was from Virginia, one of the Confederate States, and Wilcox was from New York, on the Union side. Gibbs lost almost all of his money during the war and legend has it that after the war ended in 1865, he borrowed a suit and walked 400 miles to New York City to reunite with Wilcox, who had done the decent thing and retained Gibbs's stake in the company. The company went from strength to strength after that, and from 1889 they started producing lockstitch machines as well as chainstitch ones. By the first decade of the 20th century, Wilcox & Gibbs had branch offices in major cities including Manchester, Leeds, Glasgow, Paris, Milan, Brussels and Belfast, complete with repairing departments. Brown & Sharpe continued to manufacture the Wilcox & Gibbs chainstitch sewing machine (in its original shape) until 1948.

Early Wilcox & Gibbs machines were made of cast iron and, rather beautifully, featured an arching shape that formed the letter "G" for Gibbs. I bought a hand-cranked Wilcox & Gibbs sewing machine, complete with its lovely "G" shape, which I found on eBay. I chose it because it was very similar to one I'd repaired on *The Repair Shop* a few years ago, and I fell in love with that machine. Like the machine I restored, it was in terrible shape when I got it (not quite as bad as the one I repaired on the show), but because I'd already done one successful repair, I knew what I'd got myself into. This was a labour of love, and wow, was it worth it, unearthing beautiful gold floral transfers on the wheel. And I wasn't the only one on *The Repair Shop* to fall in love with the Wilcox & Gibbs sewing machine that we fixed up: Jay Blades bought one too, and of course that ended up with me having to fix it up. Sometimes it seems like I run a repair shop within *The Repair Shop*, but I love doing this kind of stuff. And knowing that it's going to someone who not only absolutely loves it, but will use it as well, is fantastic.

One incredible thing I discovered while doing the research for this book is where my Wilcox & Gibbs machine was made. It turns out that some of the models Wilcox & Gibbs made for the European market were slightly different to the US versions. Although the machine heads were made in the US and shipped over, some of the hand-cranked wheels and mountings were cast at the Coalbrookdale works in Shropshire's Ironbridge Gorge (a UNESCO World Heritage Site thanks to its incredible contribution to the birth of the Industrial Revolution). I've been to the bridge several times – Jay from *The Repair Shop* lives close by and has a workshop there. The bridge itself is an industrial masterpiece, made of modular cast-iron sections all riveted together. These sections were made at the Bedlam Furnace, a stone's throw away from the bridge itself and you can still see the remains of the furnaces there. Coalbrookdale became legendary across the world for its decorative ironwork in the 19th century, and I'm lucky enough to have a piece of that heritage right here in my workshop. If you've ever seen the bronze-painted cast-iron gates that divide Kensington Gardens from Hyde Park in London, you'll know what I'm talking about. (If not, go and see them.)

BY 1913, THE SINGER COMPANY WAS MAKING 3 MILLION SEWING MACHINES A YEAR, ALTHOUGH EVERYTHING WAS ABOUT TO CHANGE.

HYDRAULIC PRESS

MACHINE NUMBER 041

A machine press, which is usually just called a press, applies mechanical, pneumatic or hydraulic pressure in order to cut or shape a workpiece.

Joseph Bramah was a Yorkshireman, who moved to London after training as an apprentice carpenter and found work as a cabinetmaker. By the mid 1770s, he found himself installing water closets to wealthy clients of a Mr Allen; and had a patent granted in 1778 for a hinged valve under the pan that sealed the bowl, guarding against the effects of freezing temperatures. He started producing these new water closets from his workshop in Denmark Street in London's West End and the company he founded was still going strong in the 19th century. Apparently there's a still-working Bramah water closet in Osbourne House – Queen Victoria's Italianate residence on the Isle of Wight.

In the early 1780s, Bramah developed a fascination for devising a pick-proof lock. In what sounds like a clever bit of marketing, he created a huge "challenge" padlock in 1784 (the same year he set up his Bramah Locks Company), offering a prize of 200 guineas (equivalent to more than £25,000 today) to anyone who could create an instrument to pick the lock. This challenge ended up going on until the Great Exhibition of 1851, 67 years later, when the American locksmith Alfred Charles Hobbs managed to pick it, although it took him 16 days to do so. The lock is now in the Science Museum in London.

Bramah's work as a locksmith meant that he developed many different machine tools, the most famous of which was the hydraulic press, which he patented in 1795 (it might well have been inspired by the time he spent installing toilets). Bramah's hydraulic press made use of Pascal's Law – the principle that pressure within an enclosed liquid is transmitted equally in all directions. Bramah's press featured two cylinders and pistons of different cross-sectional areas and worked like a kind of mechanical lever – applying a force to the smaller cylinder magnified the effect at the larger cylinder.

Harnessing hydraulic power was a major milestone in the history of machinery. Anything that involves heavy lifting, pushing or pulling can be made much simpler using hydraulics. And they're used in everything from the jack in the boot of your car to a space rocket launch.

Bramah worked closely with Henry Maudslay, a genius inventor and tool maker in his own right (see page 78). Maudslay worked with Bramah from the age of 18 and later went on to set up his own business.

I've got a hydraulic press in my workshop and to be perfectly honest, it's dreadful. It's too small and keeps leaking, so I have to top it up all the time, but I'd be lost without it and use it all the time. It does the job… just. And for that reason, it's not going anywhere.

The chap that moves the Ranalahs for me (see page 104) with his Hiab loader crane sent me a picture one day of an ancient hydraulic press with the caption "Dom – do you want this?!" He was delivering something to a workshop and, in a completely overgrown part of a yard was a massive 80-or-so tonne (89-ton) hydraulic press that had been living outside for years. It was so covered in brambles you could barely even see it. The guys there said he could have it and all he wanted from me was the cost of the fuel for delivery, so obviously I said yes. He put a strap around it and broke it free of its prickly prison and dropped it off at *The Repair Shop* barn. Just as well that he did, because I've got no room for it in the workshop. Fixing that one up will be a labour of love, but I'll get there. I always do.

THE HYDRAULIC PRESS
Everything that takes a lot of heavy lifting, or pushing, or pulling benefits from hydraulic power – whether that's a car jack or a rocket launcher.

RIVET-MAKING MACHINE

MACHINE 042

A rivet is a thick metal pin that passes through pre-shaped holes in two or more metal plates (or other materials including wood, clay and fabric) in order to hold them together. It typically has a mushroom-shaped head, while the other end, the tail, is beaten with a hammer until it expands so it looks much like the shape of the head end, sealing the gap. Unlike bolts and screws, rivets are designed to be permanent fasteners. So once they're in place, they're not going anywhere unless they're subjected to unbelievable force.

Although rivets are probably most associated with iron and steel buildings, ships and vehicles of the 19th and 20th century, they actually date back to Ancient Egypt, where they were used from at least 3000 BCE to fix handles to jars. They feature in some of the most iconic structures on Earth – the Golden Gate Bridge (1.2 million rivets), the Eiffel Tower (2.5 million rivets) and the RMS *Titanic*.

The *Titanic* is estimated to have contained more than 3 million rivets to keep the thousands of metal plates in place on the decks and sides of the ship. Four- or five-man riveting teams, each with specific jobs, went to work. First there was the rivet heater, who heated the rivet to over 1,000°C (1,832°F) until it was red hot. Then, in an action that would have anyone working in health and safety collapse, he removed the rivet using a long pair of tongs and lobbed the red-hot rivet over to the rivet-set holder who caught it using a tin scoop. The distance between the two men could sometimes be as much as 15m (50ft). The catcher used a short-handled pair of tongs to pick up the rivet, push it into the hole, and clasp it by the tail. At which point, the third man – the "holder up" – used a 6kg (14lb) hammer to keep the rivet in place from the other side so that the final part ot the team,

the riveter, could hammer the tail of the rivet into shape. Riveters sometimes worked in pairs, swinging hammers weighing between 1.3 to 2.3kg (3 to 5lb) to round over the rivet tail. When the metal cools, it contracts and seals the gap. You can imagine how loud and hot the working environment of a riveting team must have been.

In 1998, Tim Foecke, a metallurgist working for the National Institute of Standards and Technology in the US, analysed the wrought-iron rivets recovered from the *Titanic* and found that they were made of low-quality iron (containing a lot of slag). This meant they would have been more brittle in low temperatures, causing the rivet heads to break off after the impact of the iceberg, which let the icy water in between the hull plates. Sonar mapping of the hull of the ship has revealed much smaller and thinner tears in the hull caused by the iceberg than previously thought, and these, combined with the weakness of the rivets, probably caused the ship to sink as quickly as it did.

All rivets were made by hand, until French boilermaker Antoine Durenne invented a rivet-making machine in 1836, adapted from a punching tool he'd invented. His rivet-making machine was a multifunctional contraption, also able to make bolts and nails. At the London World's Fair of 1862, the entrepreneur, engineer and inventor Charles de Bergue exhibited a machine that could produce 2,000 rivets per hour, which really heralded the age of rivet mass production. Rivets were the dominant fasteners in all types of construction from the 1840s until the 1930s, when welding became the quickest and most cost-effective solution.

Whenever something comes into the barn for me at *The Repair Shop* which has been riveted together, I know I'm going to be in for a challenge: rivets and riveting are geared towards fixing something in place permanently. So first you have to grind the mushroom-shaped head of the rivet off (without damaging any of the surrounding area), then you drill a hole through the centre of the rivet, which releases the pressure and allows it to shrink. Then you can wiggle it out.

RIVETING MACHINE

MACHINE
043
NUMBER

A riveting machine automatically inserts and deforms rivets to join two surfaces together. One of the first is thought to have been made by a Mr Garforth around 1847 to rivet boilers. It sounds like a variant of a steam hammer (an industrial hammer driven by steam).

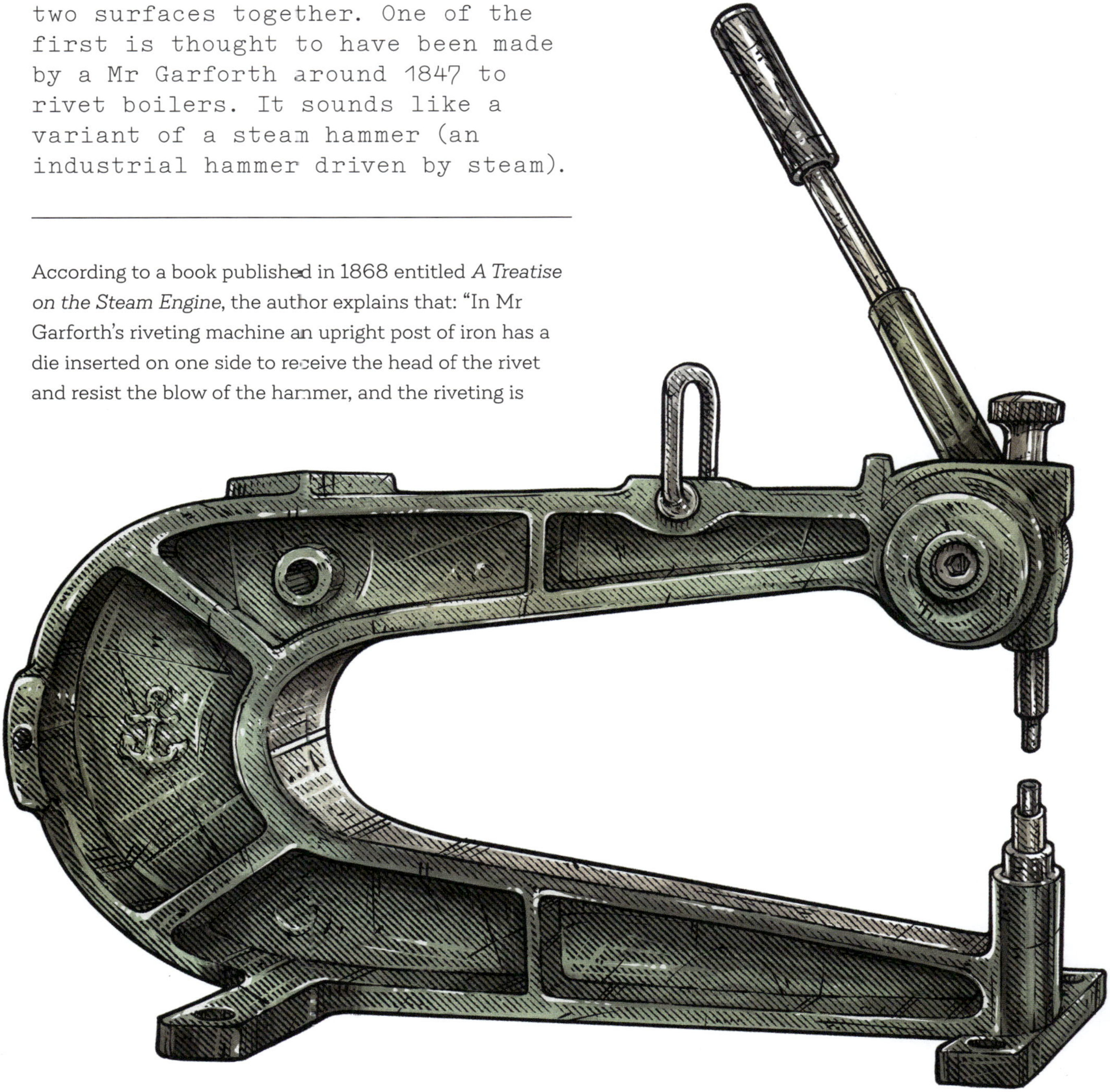

According to a book published in 1868 entitled *A Treatise on the Steam Engine*, the author explains that: "In Mr Garforth's riveting machine an upright post of iron has a die inserted on one side to receive the head of the rivet and resist the blow of the hammer, and the riveting is effected by a die attached to the piston rod of a short horizontal cylinder of about 2 feet in diameter, behind which steam is admitted."

His invention got a glowing review in the catalogue for the Great Exhibition of 1851 (the one that Prince Albert organised, which was held in the Crystal Palace in London's Hyde Park): "With this machine, one man and three boys can rivet with perfect ease, and in the firmest manner, at the rate of six rivets per minute, or three hundred and sixty per hour."

Steam-powered riveting machines had only a relatively brief time to shine because from around 1860, riveters were being powered by something simpler, more reliable and more effective: hydraulics. But hydraulic riveters were still massive machines. And then, in 1871, engineer Ralph Hart Tweddell designed a portable hydraulic riveter, which was manufactured by Fielding & Platt in Gloucester, UK. This was a massive moment because it meant that you could bring the riveting machine to the job, which was usually huge in scale, like a locomotive or a bridge. Tweddell's riveter was used to rivet a lattice girder bridge approaching Bishopsgate Railway Station in London in 1873. After that proved successful, Tweddell's invention was in constant demand across the UK and Europe.

PORTABLE HYDRAULIC RIVETER
Once a portable riveting machine was designed, the machine could be taken to the job which was quite a game-changer.

Ranalah, the company that made some of the finest wheeling machines in history, also made a riveting machine in the 1940s. This was used to make some of the legendary British fighter planes in the Second World War, including Spitfires and Hurricanes. But aside from an entry in an old trade catalogue from the 1940s, I've never actually seen one. It was advertised in the catalogue as being "specially designed for working aluminium" and promised "foolproof operation, even with unskilled labour" and that there was "nothing to wear out owing to simplicity of design". Who knows – maybe someone reading this might be able to point me towards one in an old workshop or garden somewhere.

TWEDDELL'S RIVETER WAS USED TO RIVET A LATTICE GIRDER BRIDGE APPROACHING BISHOPSGATE RAILWAY STATION IN LONDON IN 1873.

LETTERPRESS

MACHINE NUMBER 044

```
Letterpress is the oldest traditional printing technique
that there is; it involves pressing an inked, raised
surface onto paper. Every letterpress printer knows the
name Adana, the British company founded in 1922 by
Donald Aspinall.
```

Aspinall, a keen hobbyist printer, volunteered to fight in the First World War but he must have lied about his age because he was 17 at the time and you could only serve from the age of 18. He was discharged from the Army suffering from shell shock after a few months, but struggled to find work back home. So he designed a small flatbed press and advertised it in the magazine *Model Engineer*. Suddenly he was receiving shedloads of orders and cheques in the post. He couldn't possibly keep up with demand, so he sought the advice of a policeman at Twickenham Police Station in south west London, and the story goes that he advised Aspinall to "make the presses!". Four years later, Aspinall founded the Adana Agency, selling small, reasonably priced flatbed wooden presses to hobbyists. For a while everything was going really well and Adana was exporting its presses to the US and Australia. But it overreached itself, and by 1939 the company (and Aspinall himself) was in poor health, eventually being bought cheaply by one of its creditors, Frederick Ayers. Like many non-essential businesses in the Second World War, it basically went dormant, but did end up producing presses for the resistance movement in various parts of Nazi-occupied Europe.

It pressed on after the war, and by 1952, you could pick up an Adana in any one of 100 countries around the world. The following year it produced its most famous model, the "Eight-Five", named for the dimensions (in inches) of the chase – the metal frame used to hold the type. Two further variants were manufactured – the Mk II from the early 1960s and the Mk III from the early 1970s.

Adana's letterpresses were built to last and they're also impressive-looking machines. I used an Adana Eight-Five when I was at art college. I have very fond memories of my Adana because my first business cards (some of which I still have) were printed on it.

Adana suffered during the 1980s as computers came to the fore, and the company was bought by Caslon Limited in 1987. You might recognise the name Caslon because it's one of the most famous typefaces in history, named for the company's founder, William Caslon, a legendary type founder. Its most famous use was for the printing of the US Declaration of Independence in 1776.

Letterpress made a big comeback in the 2010s, and Caslon produced a new version of the Eight-Five, the Adana 85C, in 2017. And that's something a passionate advocate of heritage craft like myself absolutely loves to hear.

THE ADANA EIGHT-FIVE
Built to last, the Adana Eight-Five was named after the dimensions (in inches) of the metal frame area used to hold the type in place.

HE DESIGNED A SMALL FLATBED PRESS AND ADVERTISED IT IN THE MAGAZINE *MODEL ENGINEER*. SUDDENLY HE WAS RECEIVING SHEDLOADS OF ORDERS.

BRILLIANT-CUT MACHINE

Brilliant cutting is the technical term for hand-engraving patterns into glass using a rotating stone wheel. It became popular in the 1890s and you can still see the amazing results of brilliant-cut panels and mirrors in pubs dating back to that time.

London's West End has the highest concentration of these features, but you can also see them in Leeds, Manchester, Birmingham, Brighton, Liverpool and Glasgow and in country pubs across the land. Historically, brilliant cutting costs more than embossing (which involved creating decorative designs by applying acid to a template) but it's unmatched as a technique because the numbers and letters stand out incredibly well. The rotating wheel cuts a groove into a template drawn on the glass and the brilliant cutter moves the glass towards the wheel to control the depth and shape of the cut. Sometimes they had to use a harness to hold the glass up because controlling the machine and holding the glass at the same time wasn't half hard work. Because they were cutting at an angle, it created a dazzling appearance when light passed through it. The design was polished and could be embellished with gilding – adding gold leaf to colour in the grooves. A lot of the designs from that era were in the Art Nouveau style and so feature ornate symmetrical floral patterns combined with stunning lettering.

It's a seriously labour-intensive technique that requires skill and patience, and the great news is that there has been a revival in brilliant cutting. There are a handful of craftspeople leading the charge and I'm lucky enough to know one of the best signwriters, brilliant cutters and traditional glass artists in the UK, if not the world: David A. Smith, MBE. Dave's work is utterly incredible. I went to a brilliant-cutting group course he offered down in Devon years before *The Repair Shop* and it was absolutely mind-blowing. He's got the most amazing collection of old glass signs and just getting the chance to talk to him and see him work was really inspiring.

As for brilliant-cut machines, they are one of the very few machines in the book that I don't think have ever been commercially made. They always look like something someone has put together in a shed using odd ends of wood, half a scaffolding pole and a bunch of clamps. It's an amazingly ramshackle operation for something that requires such precision, but that's what I love about it. You don't need the latest and greatest machines to create the most incredible pieces of work. In this case, you just need a structure to support the turning wheel so that it runs true. The wheels are the amazing part of the machine – Dave's got a whole wall of different sized and angled wheels to match whatever he's cutting, and the whole set has been passed down from brilliant cutters for generations before him. You can't buy new ones anymore, so they're really precious items. The wheel rotates through a washing-up bowl (or Tupperware or old margarine pot) containing water and a wet rag to stop the wheel and glass from heating up and to lubricate the wheel as it's turning. I've had a go on Dave's machine, and I really got a sense of the precision and dedication you need to excel at glass cutting. Some pieces of glass artistry take weeks if not months. And you're only ever one wrong move away from a disaster when you're working with something as fragile as glass.

I've been promising Dave I'm going to make him a new machine, but he's perfectly happy with his setup now. And there's something uniquely magical and personal about having to make your own machine in order to become a brilliant-cutting expert.

BRILLIANT-CUT MACHINE
There have never been commercially produced brilliant-cut machines. They generally look like something someone has cobbled together in a shed.

WHEELING MACHINE

MACHINE NUMBER 046

The wheeling machine (known in the US as the English wheel) is a manually operated metalworking machine used to create curves in flat sheets of metal. An early version may have been used to make armour in France during the Middle Ages but they were starting to be used widely in manufacturing only in the late 19th century. Their heyday, however, was the early 20th century when the wheeling machine proved to be the perfect way of shaping panels for cars, buses and aeroplanes.

From afar, a wheeling machine looks like a massive C-shaped clamp, but the business end is at the points of the "C" where there are two wheels – a larger one on the top (the rolling wheel), and a smaller one on the bottom (the anvil wheel). Both the wheels are interchangeable, but it's mostly the bottom one that you change depending on the stage of the curving process you're at and the curved finish you're looking for. The throat of the machine is deep to allow the user to manoeuvre large sheets of metal, and you can adjust the gap between the two wheels depending on the thickness of the metal sheet you're using. You feed the sheet metal between the two wheels which stretches and thins the metal. It's a bit like a rolling hammer, if you can imagine that.

My first experience with a wheeling machine was going on a week-long course run by Geoff Moss at MPH Motor Panels in Cornwall, back in 2020. I had the best time down there, I really did. His working set-up is about as analogue as it gets. There isn't a single power tool in there, and if you're looking for a plug socket, you're best off trying next door. He's a brilliant guy and an absolute genius at working with wheeling machines. When I first visited him, he had two wheeling machines, one made by F. J. Edwards and the other made by a company called Ranalah, which I learned was widely considered to be the Rolls-Royce of wheeling machines. After Geoff had armed me with the basic skills, I bought myself a wheeling machine made by F. J. Edwards, and became completely addicted. The skill, know-how and rhythm you need to use the machine all hugely appealed to me. It was just such a magical combination. Seeing Geoff form a compound curve was like watching an artist at work. Everything was so smooth and finely honed – testament to decades of essentially becoming an extension of the wheeling machine. I wanted to become that good. And I wanted the best machine there was. I wanted a Ranalah. But, try as I might, I just couldn't get hold of one.

The Ranalah story begins with two coachbuilders – John I. Dalrymple and Charles H. Livesay – who set up a company called John Charles & Company in 1932 near Kew Gardens, west London. John Charles & Company made the bodywork for a whole range of cars including the Ford Eight, Lagonda Rapier and Morris 10. It started using the brand name Ranalah on car bodies from around 1934 and began to develop an international reputation for the quality of its metalwork. The company moved into a vast new office in Brentford, west London, but overreached itself and was sold in 1935. A new company – Ranalah Coachworks Limited – took over the assets of John Charles & Company and moved operations to Morden Road in Merton, south London. In 1937, Ranalah was contracted by the UK government to make panels, tubing and other metal parts for aircraft including the legendary Supermarine Spitfire, which, along with the workhorse Hawker Hurricane, helped to win the Battle of Britain in 1940. Soon after winning the government contract, Ranalah started making its own wheeling machines because the other machines on the market had a straight leg at the front which a large panel for an aircraft would bang into. That's why the Ranalah wheeling machine is the unique shape it is. It was designed to make a Spitfire. And that's an incredible thought. Its shape was a functional design, but Ranalah chose to make it beautiful. You can tell it was made by a coachbuilder with an eye for design and detail.

After the war, when the government was trying to recoup money, it sent Ranalah a big backdated tax bill which Ranalah couldn't afford to pay. The owners had no choice but to liquidate the company. The assets were sold and eventually absorbed into another larger company, so the name Ranalah faded away, except for those with a specific interest in cars and aircraft of the 1930s and 1940s. I'm absolutely one of those people, and I fell in love with the company. So after following the paper trail, which took me to an industrial estate on the edge of Brighton, then hiring a lawyer and filling out all the documentation, I became the very proud owner of Ranalah Limited in 2021. I just couldn't let all that proud history fizzle out.

My idea was to bring Ranalah back, start producing its wheeling machines again, to the exact specifications of the originals, and try to get the younger generation passionate about wheeling.

My first job was to get hold of an original Ranalah cast-iron wheeling machine, so I could make an exact wooden replica. Then I'd take that to a foundry I'd got lined up – East Coast Casting, one of the only foundries in England that could cast something weighing 750kg (1,653lb) and with dimensions of roughly 2m by 1.2m (6ft by 4ft). There, they press the wooden model into a bed of sand, which leaves an imprint so that they can then pour in the molten iron. After that has cooled down, you've got your cast-iron frame. I've missed out a few stages, but that's the gist of it. I had a plan, and I had one big problem – I couldn't get hold of an original Ranalah.

I did a shout-out on YouTube and Instagram, but there were no leads. And then, weeks later, out of the blue, I got a message from someone on Instagram with a photo attached. It was a picture of a wheeling machine almost completely covered in ivy and brambles in the middle of some woods. "Is this what you're after, Dom?", the message read. Unfortunately though, it wasn't his photo. He'd heard my shout-out and had found the picture of the wheeling machine on a Facebook group. So I had to find the original message, ask the admins if I could join the group, and then find

MY IDEA WAS TO BRING RANALAH BACK AND START PRODUCING ITS WHEELING MACHINES AGAIN, TO THE EXACT SPECIFICATIONS OF THE ORIGINALS.

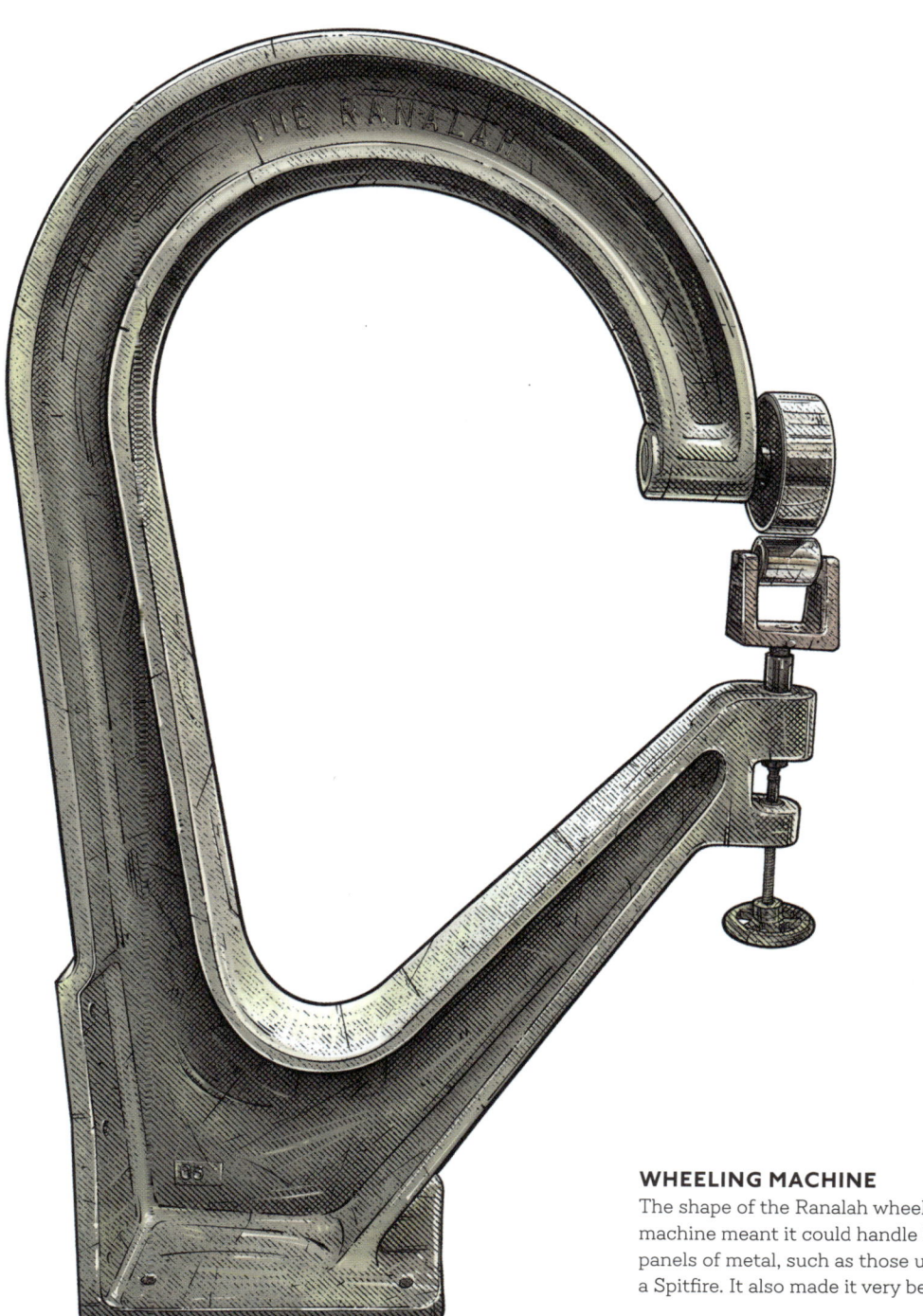

WHEELING MACHINE
The shape of the Ranalah wheeling machine meant it could handle large panels of metal, such as those used for a Spitfire. It also made it very beautiful.

the guy who had posted the photo. And I had so many questions. *Where was it? Was it still there? Was it a Ranalah?* The guy told me he'd been on a footpath between two villages near someone's house when he took the photo, and gave me the rough postcode. I spent the whole of that evening on Google Street View trying to find the exact place, but no joy. Nevertheless, I had a rough idea where it was, so I woke up the next morning and jumped straight in the van, flung the walking boots in the back, and drove for two hours to try and find this machine. The search was on.

I spent hours in the pouring rain walking around every path, track and field in the surrounding area but couldn't find the place. Eventually, I walked across some farmland and got accosted by an angry farmer, but I explained what I was doing and his expression changed. He knew exactly what I was talking about. A ray of light. He pointed me to a house nearby, although he warned me that the guy who owned the Ranalah had just moved out, so the Ranalah was most likely gone. I had a snoop around and finally found the site from the photograph, only to see cut pieces of ivy and the imprint of a wheeling machine on the ground. It had been moved very recently. I was gutted.

But then, just as I was about to drive home, a chap came out of his house to talk to me. He told me that the guy who owned the wheeling machine was an elderly former engineer who'd just moved in with his daughter, and he gave me her number. I rang her, asked about the wheeling machine, and she said, with more than a touch of relief, "That old wheeling machine is stuck on my driveway. I'd love to get rid of it!" I couldn't believe my luck.

I got to the address and there it was. A wheeling machine sitting on its side, on a pallet, in the mud. And when I got closer, I saw those magical words, "The Ranalah" on the frame. I could have cried. And then the elderly owner came out – what an incredible chap – and he beckoned me to come and check out his car collection. You don't need to ask me twice. There was a 1920s Rolls Royce, a turquoise 1930s Bentley and a 1930s 2.5-litre red Maserati, which he'd made himself based on an original model he owned. And he'd used his Ranalah to make the bodywork. He was an absolute genius of a man who put my skills to shame. And he agreed to sell his Ranalah to me. A week later, I came back with a guy I'd hired with a truck and an Hiab to lift it. And that was terrifying, because if the rope slipped, the Ranalah might just snap in half. But it went down fine, and made it over to my workshop. I couldn't believe the adventure I'd been on already. But I was about to start an even bigger one.

First, I needed to take apart the components and get them stripped down and cleaned so I could get wooden patterns made. Then it was on to Paul Cameron at Modular 105 to map out the exact dimensions of each component of the wheeling machine, fire up his computer numerical control (CNC) machine, and cut them to size. For the frame itself, Chris Isbill at East Coast Casting in Norfolk gave me the tip to use Michael Young at MJY Patterns to make the pattern. And he did a sterling job before transporting it back to East Coast Casting to make the frame. I watched the whole casting process, which was utterly incredible, before I saw my original Ranalah and the new "No. 1" casting next to each other, and they were a perfect match. It blew my mind that there were nearly 100 years between them.

The first Ranalah to leave my workshop (the "02") went off to Geoff in Cornwall, who's got a lifetime's experience of passing a piece of metal through an original Ranalah. It was important to me that the new Ranalah felt like the original, and there's no one who would know better than Geoff. I wanted a seal of approval from a master wheeler. And, thankfully, I got it.

I donated one of my new Ranalahs to Aston Martin because it had recently started up an apprenticeship scheme in metalworking (in collaboration with the Heritage Skills Academy, who I do ambassadorial work for), and I really appreciated what it was doing. There were two apprentices but only one wheeling machine, which is why I offered them the Ranalah. The other wheeling machine, an original Ranalah, is in the Aston Martin plant up at Newport Pagnell. Thinking about the cars it would have built, maybe even the DB5 used in *Goldfinger* or *Thunderball,* just gives you goosebumps. And that association with Aston Martin is the reason why I numbered the seventh Ranalah I've had cast as "007" for Goodwood Revival.

While all this casting was taking place in spring 2023, my friend Steve Fletcher made an exciting Ranalah discovery. He found what turned out to be a "Type M" Ranalah wheeling machine in Witney, Oxfordshire. It was one of the other two types of wheeling machine Ranalah made and was throatless and designed to be bolted to the floor. It would have been used to make big aircraft panels, most probably for de Havilland, which had an airstrip nearby. It might well have been used to make the few metal panels in the famous de Havilland Mosquito, first introduced in 1941. I was so excited about getting that machine back to the workshop, and trying it out. It also made me imagine the guys who originally bought it from Ranalah and installed it, getting really excited about using it for the first time. That's the kind of passion I want to instill in the younger generation. I want to spend time going to schools, colleges and academies, like the old salesmen used to do back in the 1950s with their scale models of machines, to get potential customers excited about trying them out. In that vein, I'm in the middle of experimenting with a (slightly under) half-size version of the Ranalah so it is more affordable and more accessible, and doesn't take up as much room. Wheeling machines are remarkable objects with a remarkable history and I'm hugely passionate about seeing the craft thrive again.

As a final thought, I spent a lot of time wondering about the origin of the name Ranalah. I've also had a lot of folks ask me about it. So I did a bit of digging, and I think the most likely origin of the name comes from an alternative (and less posh) spelling of Ranelagh. I think it was used in homage, and possibly even as an in-joke, between John and Charles, the two lads who formed John Charles & Company in 1931. John and Charles had both been employed by coachbuilders Chalmer & Hoyer sometime after 1924 when the company opened a factory in Weybridge, Surrey. They must have been from the local area, given that people didn't tend to travel far to work back then. And seeing as they were both similar ages, I think it's very likely they became good mates there.

The first car manufacturer to use the name Ranelagh was Standard with its model "S" car, but it wasn't exactly an inspiring car that two lads might pay homage to. But the second car maker to use the name Ranelagh was Austin, which named its two-door, coupé version of the Austin Twenty the Ranelagh in 1921. The Twenty won lots of admirers after its performances at Brooklands (incidentally, very close to Weybridge) in front of large crowds. So it's possible a young John or Charles, or both, saw an Austin Twenty Ranelagh at Brooklands, fell in love with it and then, years later, came up with the name Ranalah in tribute.

There's one final possibility: the next time Ranelagh was used by Austin was for a top of the range, six-cylinder, limousine version of the Twenty, which was announced at the London Motor Show in 1926. Given that this was considered equal to the 20hp Rolls-Royce, this might have been the most desirable car out there for two lads who spent their entire time at Chalmer & Hoyer putting together boring Morris Oxfords. You've got to dream, haven't you?

Then, just a few years later, John and Charles had formed their own company and were building the bodywork for Rolls-Royce, Bentley and Crossley. So I think Ranalah probably reminded them each day of a dream they had had when they were younger. All this is guesswork, but if anyone reading this can shed any further light, do get in touch.

SCAIF

MACHINE NUMBER 047

The scaif is a rotating polishing wheel used to polish diamonds. It was invented in 1456 by the Flemish jeweller and diamond polisher Lodewyk van Bercken and completely transformed the way diamonds were shaped.

With his invention, he could smooth out the rough surfaces of diamonds using an abrasive lubricant made from mixing olive oil and diamond dust. The use of diamond dust was important because diamonds are so hard that the only object able to cut them was another diamond. That isn't so anymore, with the advent of modern laser-cutting techniques and diamond saws.

The scaif is still used today in diamond polishing. It rotates the same way as a potter's wheel and the top surface of the disc is covered with diamond powder. The diamond is gripped in a specialised clamp and each of the diamond's facets are polished, one after the

SCAIF
Diamond surfaces are so hard that they are polished using diamond dust. This happens on a scaif, a machine that was invented in 1456 and is still in use today.

other. Many different techniques are used before diamonds can be polished, though. The rough diamond is usually cleaved using another diamond (or a laser) or sawn using a circular steel blade (lubricated with oil and diamond dust) to split the stone into smaller pieces. This can take days to achieve. Then the diamond undergoes a process called bruting: where the diamond to be shaped is set onto a spinning axle to grind against another diamond turning in the opposite direction. This creates the diamond's facets and it's seriously intricate work. Next the scaif symmetrically polishes the diamond's facets and smooths out any remaining roughness.

PHARMACIST'S PILL ROLLER

MACHINE NUMBER 048

If you've ever wondered how medicinal pills were originally shaped, they were rolled using a specialised two-piece pill-rolling machine. This was a sloping base, with two brass side rails, and a fluted-brass rolling platform with 24 individual channels.

The second part of the machine, the roller handle, had four wheels in the side rail and a matching fluted brass section. The chemist first prepared the mixture into quite a stiff mass before rolling it into a sausage shape, which they stretched across the width of the 24 grooves. Then, they used the roller handle to cut the sausage shape and roll the pieces out. Next, the pieces were moistened using gum Arabic (a naturally occurring gum originally imported from North Africa) or mucilage (another sticky gum) and placed inside a pill silverer. It was once common practice to coat medicinal pills in silver or gold leaf, to cater for the fine tastes of the wealthy who could afford these types of medications. Pill silverers were usually globe-shaped wooden boxes with a domed lid on a screw thread and a very smooth inner surface lined with silver leaf or gold leaf. Once the pills had been placed inside, the top was screwed back on and the whole thing rotated for a few minutes to coat them in the leaf. And finally, they were placed in a pill box, which could be very elaborately decorated and made of sterling silver or gold, ready for the customer. No expense was spared back then.

Pill rollers are beautiful objects, usually made from mahogany and brass, and would have sat on the chemist's countertop. One well-known company that produced them during the 19th century was S. Maw, Son, and Thompson (originally founded by George Maw in 1807), based on Aldersgate in the City of London. They made a range of chemist equipment, surgical appliances, medical instruments and pharmaceutical products. They also produced some slightly more unusual items, such as the "perfume pistol", which dispensed perfume from a glass barrel. It didn't take off.

PHARMACIST'S PILL ROLLER
These beautiful objects, often made of mahogany and brass, used to sit on the chemist's countertop.

IT WAS ONCE COMMON PRACTICE TO COAT MEDICINAL PILLS IN SILVER OR GOLD LEAF, TO CATER FOR THE FINE TASTES OF THE WEALTHY.

CIRCLE-CUTTING MACHINE

MACHINE NUMBER 049

I've got an old cast-iron manual circle-cutting machine in my workshop (I also take it to my little blacksmith's barn at *The Repair Shop*) and it's an incredible-looking machine.

It's got two sections to it: the clamping jaw, which holds down the centre of the piece of metal you want to cut, and the crank handle which turns two circular blades (above and below your metal sheet) and pulls the metal through. It works a lot like a giant can opener.

My circle cutter was made by F. J. Edwards of London, which was one of the main companies producing sheet-metal machinery from the 1930s. I've also got one of their guillotines for cutting sheet metal and of course my first wheeling machine was made my these guys.

F. J. Edwards actually started as the Belgian Specialties Company, in 1916, but changed its name the following year and used the brand name "Besco". It started out making machinery in the cellars below St Pancras Station in London (which had historically been used almost exclusively by brewers until that point) before it moved to premises on Euston Road.

I love how complex and well-engineered this machine is just to perform one very specific task. And its other great beauty is that there's no way you can cut out a circle by hand with anywhere near as much precision. I'd been cutting out circles for years with tin snips, and so the circle-cutting machine was a godsend.

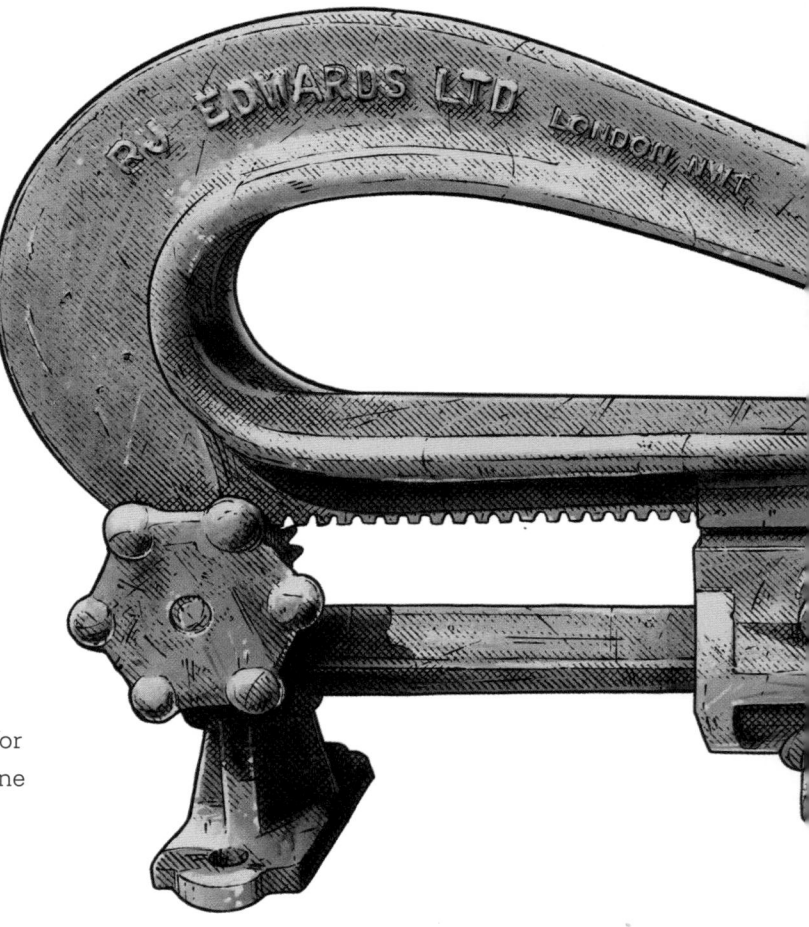

ITS OTHER GREAT BEAUTY IS THAT THERE'S NO WAY YOU CAN CUT OUT A CIRCLE BY HAND WITH ANYWHERE NEAR AS MUCH PRECISION.

CIRCLE-CUTTING MACHINE
It works a bit like a giant, beautifully engineered can opener – but one that cuts perfect circles.

ENGINE CRANE

This is a machine dear to my heart. It's a small cantilever crane designed to lift engines out of and into cars, vans and anything else with wheels. They're usually mounted on castors so that after they've lifted the engine out, you can move them away from the vehicle.

I think every garage I've worked at has had an engine crane with an "HF" stamp – for Harvey Frost. The company goes back to around the turn of the 20th century and it became very well known for its engine cranes and breakdown cranes mounted on the back of Land Rover Defenders and small trucks. I've got a Harvey Frost engine crane outside the front of my workshop, and it's a monster. It must have been designed for trucks or light aircraft because it's bigger than any engine crane I've ever seen. The elbow (that's the pivot point) of the crane is on a screw that goes up and down, which is operated with a big wheel on the back of the crane. I need to lower the elbow so I can get engines in and out of the back of my van.

My engine crane had been sitting in a field for the best part of 20 years, the base buried thigh-deep in mud, before I gave it a home in May 2022. It was in bad shape. Every part of it had seized up, so I had a job just getting it working. Usually a bit of heat and persuasion frees most things up, but this was a fight the whole way down. I had the hydraulic arms rebuilt, using all the original bits, while I dealt with the main framework, replaced the castors, and made sure everything was working as it should do. I decided not to restore it, though, so it's still rusty in places, and has moss on it, but it's made of such thick steel that surface issues aren't going to be a problem. Plus, I like machines that look like they've lived a life, and make you wonder about all the things they've been used to lift and fix. I use it almost every single day, I wouldn't have been able to lift my Ranalah wheeling machines without it, and it's an awesome piece of British engineering. It feels like I'm cheating, though, when I'm pumping a lever with one hand to pick up a tonne of cast iron. Hydraulics still feel like a magic trick. And when I crank that lever, I always wonder about the moment in history the first hydraulic machinery was tested and worked. What a day that must have been.

ENGINE CRANE
The engine crane is a machine with a very specific job: to lift engines in and out of vehicles. They are often on castors so they can be moved out of the way afterwards.

CORK PRESS

This isn't the machine used to press a cork into a bottle – we're going a step further back than that. I'm more interested in the machine used to press and mould corks before they're inserted into bottles to create a waterproof seal.

Three different styles of cork press were designed: the rotary cork press, which used a hand-operated lever connected to a ridged wheel to squeeze and roll the cork; a press that looked a lot like a hinged nutcracker, featuring several indented moulds for different sized corks; and a simple press that applied vertical pressure by means of a crank handle, forcing the cork down through a mould.

Some of them, especially the nutcracker style, were extraordinarily intricate, often in the shape of different animals, but especially snakes and crocodiles. I wonder if manufacturing chemists and pharmacies, who would have made their own cork presses, chose specific animals as a kind of insignia to separate them from the competition, or whether they were designed to be impressive so they'd be a talking point for customers. Either way, I love the commitment to making something beautiful as well as functional.

All of the different styles of cork press were heavy objects, typically made of cast iron and designed to be either bolted to a chemist's countertop or mounted on the floor. The rotary cork press was patented by C. L. Lochman of Pennsylvania in 1867 and manufactured by the Enterprise Manufacturing Company of Philadelphia, the folks who made a name for themselves for their household gadgets and industrial gadgets like the meat grinder (see page 58).

Humans have known about cork's properties of buoyancy and impermeability for thousands of years, and they've been used as wine and olive oil stoppers, buoys for fishing nets, and in the soles of sandals in the ancient world. Charred cork with laurel sap was even used as a "treatment" for baldness in Ancient Rome, thanks to the recommendation of the author and physician Dioscorides in the 1st century CE, but probably best not to give that a go at home.

Cork comes from the outer bark layer of *Quercus suber*, better known as the cork oak, which is found across southwest Europe and northwest Africa. Most cork comes from southern Portugal, and the harvesting of it is a sustainable process because the cork can be extracted from the tree without causing long-term damage, provided the tree is over 25 years old and of suitable thickness. It's a delicate operation, though, and requires strength and precision, so as not to damage the living inner layer of bark. I'm very fond of the cork oak trees in the grounds of Goodwood, the stately home in West Sussex, UK, which holds two famous annual events: the Goodwood Festival of Speed and the Goodwood Revival, the three-day festival celebrating iconic cars. In 2023, I was a part of their Revive & Thrive event – a celebration of heritage (and endangered) crafts. It's an incredible event and I'd urge everyone with an interest in craft, old cars, and all things vintage to go at least once. As for the cork oaks at Goodwood, they're incredible trees – they've got the kind of bark that you can't help but want to touch. Cork is such a visually beautiful and versatile material. It's a great insulator, which makes it a natural fire retardant. It's also very good for things like flooring and café tabletops because it manages to be strong, soft and water-resistant. I built many of the workbenches at *The Repair Shop* for my colleagues and I used cork as a topper for Kirsten's (our ceramics expert) because it's a got a bit of give, which is exactly what you're after when you're working with something as delicate as ceramics.

ROTARY CORK PRESS
The hand-operated lever on a rotary cork press was used to squeeze and roll the cork. Many chemists made their own corks for their bottles of medicines.

I WONDER IF MANUFACTURING CHEMISTS AND PHARMACIES CHOSE SPECIFIC ANIMALS AS A KIND OF INSIGNIA TO SEPARATE THEM FROM THE COMPETITION.

BOOKBINDING MACHINE

MACHINE NUMBER 052

David McConnell Smyth was an Irish-born inventor who moved to New York City, trained as a machinist and became acquainted with Thomas Edison. Smyth was awarded over 40 patents in his life for all sorts of different inventions, including devices for sewing cloth and shoes, but the one he's most known for is his book-sewing machine.

He was granted his first book-sewing patent in 1868, but it wasn't until 1879, when he'd produced an updated version of his machine, that things really started to take off. He was awarded a Gold Medal by the American Institute of the City of New York that year for his invention, describing it as "so important in its use or application as virtually to supplant every article or process previously used". His third version of his machine, the No. 3, completely changed bookbinding. It used a mechanism to pierce holes in the spines of signatures (booklet-like groups of pages, each formed from one large sheet of paper), and then curved needles used a single thread to sew through the fold in a signature and then sew each signature to the one behind it. This created a binding that meant pages could not fall out unless they were physically ripped out. This technique became known as "Smyth Sewing" and it's still used today. His No. 3 machine could sew together thousands of signatures an hour, and the quality of the sewing surpassed a human equivalent. One machine could perform the work of ten trained book-sewers.

The bookbinding business changed with the arrival of "perfect binding"– invented in the 1890s but really picking up steam in the 1930s – which used glue to bind together either signatures of a book, or single pages. While this was a cheap way to produce paperback books, it wasn't perfect, because the glue can deteriorate and the spine can break if it's stretched too wide. But it proved to be the end for Smyth's machine – the last one was sold in 1928. But, the company he set up in 1879 – the Smyth Manufacturing Company – is still in business (now as Smyth Machinery USA) selling supplies and equipment to the bookbinding trade.

The idea of an industrial bookbinding machine sounds a bit alien in the world we live in now, where bookbinding is very much a bespoke and highly skilled craft. But it shows the value of a machine for a certain time and place, designed to help produce large quantities of books to match demand.

Chris Shaw – one of my fellow craftspeople on *The Repair Shop* – is a bookbinder, and I'm in awe of what he does. To be honest, I'm quite envious, because it always seems like very clean work, unlike what I do, which often involves noise, heat, sparks and smoke. By contrast, Chris is there with his immaculate set of tools and beautiful marbling paper, and is patiently working by hand on very intricate details of books. It seems like the work is very peaceful, quiet and calm, and that mirrors Chris as a person. You know when you find someone who has found the perfect job for them? That's Chris Shaw. A master of his craft.

BOOKBINDING MACHINE
Today bookbinding is a specialist handicraft, but when this machine was invented to sew the pages of a books together, it worked at the speed of ten book-sewers.

CHAPTER

4

THE VICTORIAN AGE

MACHINE NUMBER	MACHINE NAME	PAGE NUMBER
053	MECHANICAL GYROSCOPE	124
054	SEISMOGRAPH	125
055	TELEGRAPH	126
056	WRISTWATCH	128
057	DENTAL DRILL	130
058	STEAM ENGINE	132
059	STEAM LOCOMOTIVE	136
060	SCREW PROPELLER	140
061	BARBER'S CHAIR	142
062	CASH REGISTER	145
063	PATERNOSTER LIFT	148
064	ESCALATOR	150
065	BRAILLE-WRITING MACHINE	152
066	MILKING MACHINE	154
067	MECHANICAL PENCIL SHARPENER	155
068	TICKET-PRINTING MACHINE AND TICKET-DATING PRESS	156
069	TIME RECORDER	158
070	CAMBRIDGE ROLLER	159
071	PUMPJACK	160
072	FORD'S MECHANICAL CONVEYOR BELT	162
073	FIDDLE DRILL	164

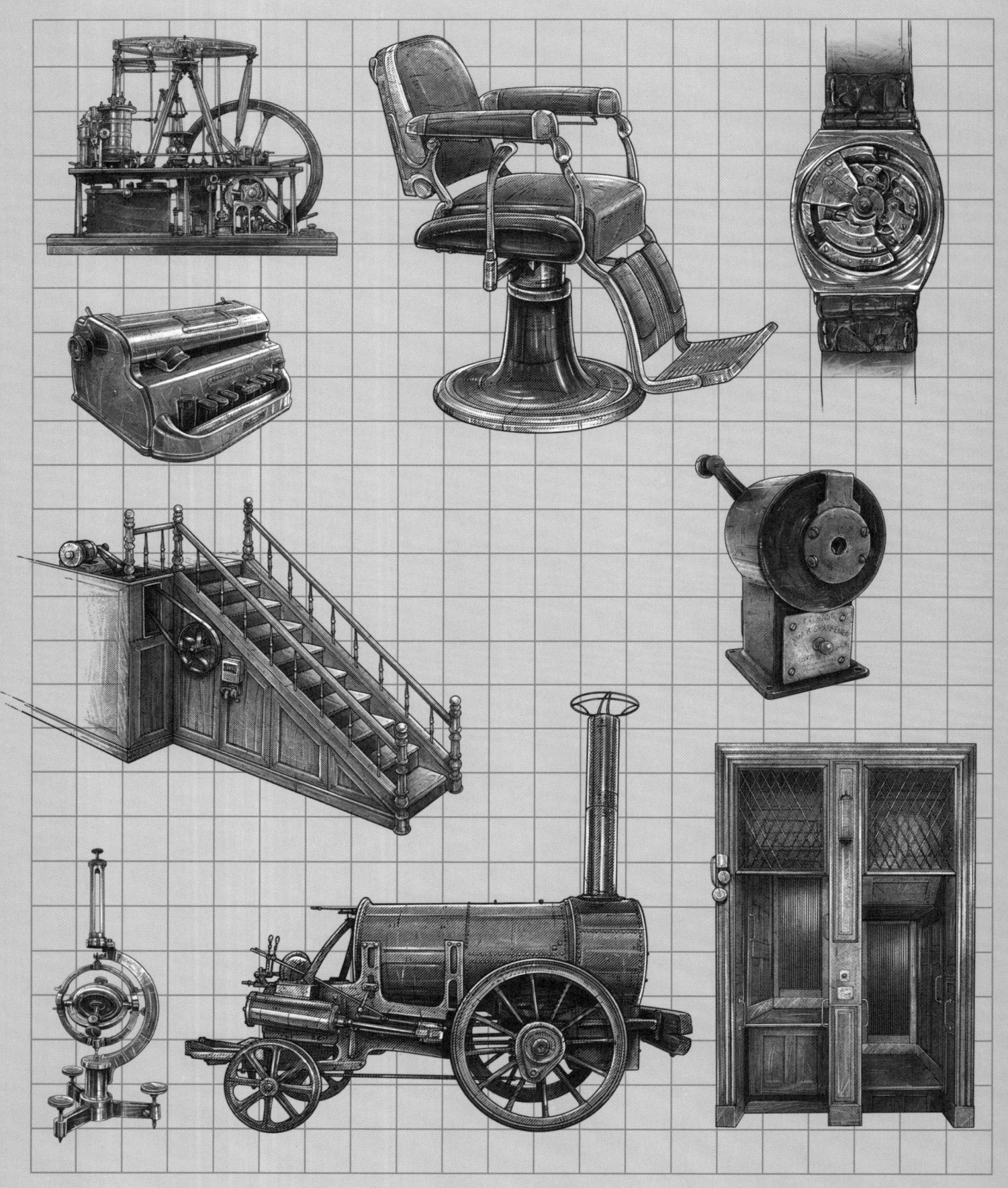

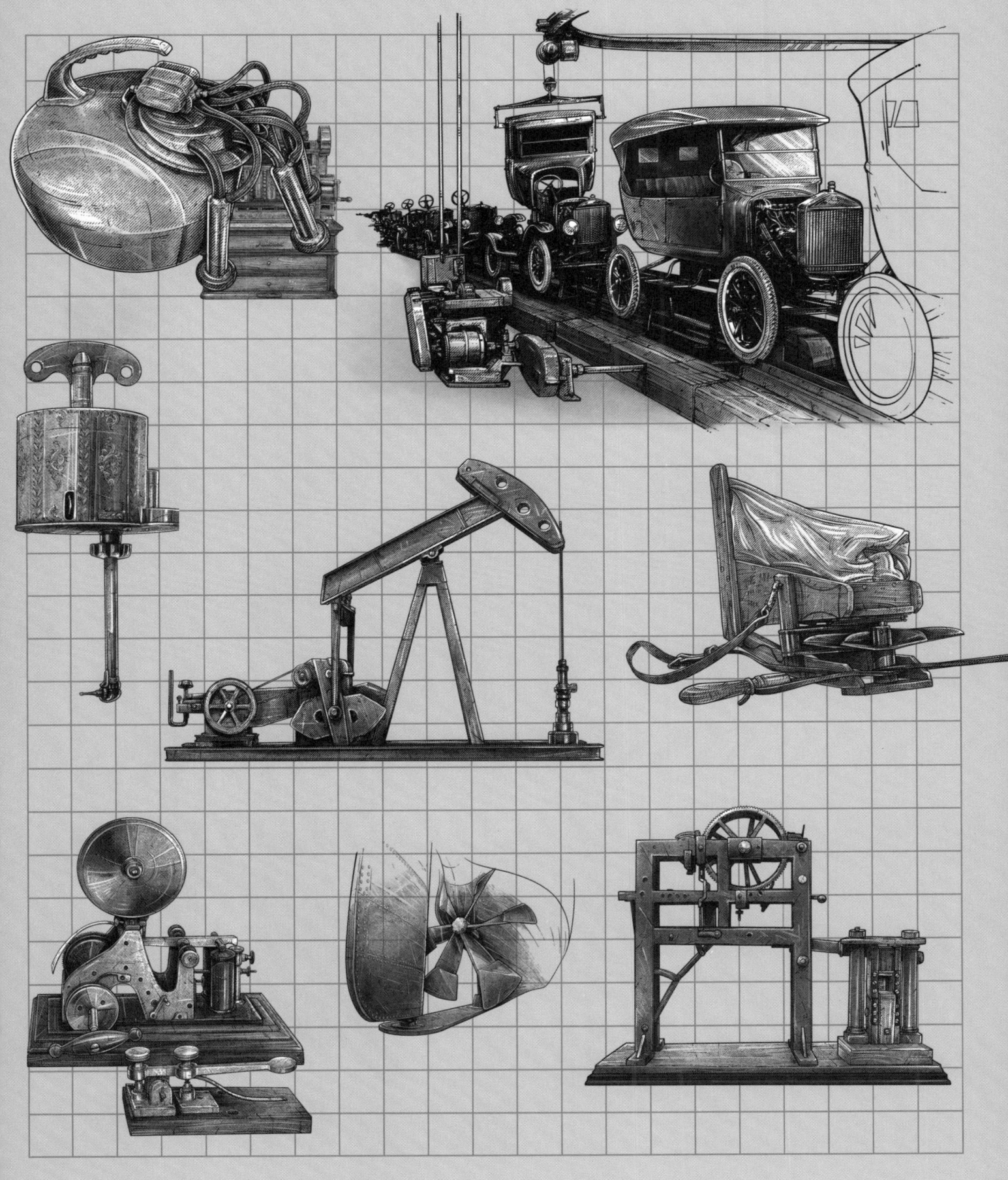

MECHANICAL GYROSCOPE

MACHINE NUMBER 053

A gyroscope is a device used principally to measure orientation. It consists of a spinning wheel on an axle, which is mounted on a pivoted circular support called a gimbal, which is connected to one or two larger surrounding circular gimbals.

When the spinning wheel turns, it creates a force (known as angular momentum), which is perpendicular to the wheel's axis of rotation. The effect is that when you rotate the whole device, the gimbals spin, but the angle of the spinning wheel is unaffected. It "wants" to continue spinning on the same axis. And there's something else that behaves in a similar way: the Earth. Our planet is basically a massive gyroscope.

The French physicist Léon Foucault created the first mechanical gyroscope (and came up with the name) in 1852, to prove that the Earth rotates on its axis. Foucault's mechanical attention to detail was spot-on, so the experiment was a terrific success. Gyroscopes have since become an essential tool in systems such as compasses, automatic pilot systems satellites, rockets, and any object that needs to know orientation.

Foucault's gyroscope is a beautiful machine. And it goes back to an era when craftspeople were inventing something for the first time and could have simplified the finishing touches but chose not to. Maybe it was to do with pride in what he was making, or knowing that it was going to be exhibited on a public stage. This was an era when your design could end up at a World's Fair, so you needed it to stand out.

It's a bit like when Steve Fletcher takes apart a clock on *The Repair Shop* and is faced with all sorts of tiny components that have been expertly and beautifully crafted with bevelled edges and engraved details. Whenever I see something like that, I'm always filled with respect for the craftsperson and how they've gone above and beyond what was necessary. Centuries later, it puts a respectful smile on your face.

SEISMOGRAPH

In 1855 an Italian academic called Luigi Palmieri became
the Director of the Vesuvius Observatory, having first
started work there in 1847.

Located on the slopes of Mount Vesuvius, near Naples, Italy, this monitoring station was tracking the activity of the volcano. And Palmieri had a job on his hands, because Vesuvius was especially active at the time, erupting five times between 1850 and 1872. Immediately before the 1855 eruption, Palmieri felt several light earthquakes. He was the kind of scientist who'd invent an instrument or machine if something wasn't doing the job he needed it to do, so he came up with a seismograph to detect ground tremors that weren't detectable by humans.

His instrument comprised a number of mercury-filled U-shaped tubes, each one pointing to a different point of the compass. When the ground began to shake, the mercury moved, closing an electrical circuit. This both stopped a clock and started moving a roll of paper. Suspended pencils traced a line along the paper tape, and these lines spiked whenever the mercury moved. So his machine could track the precise time, the duration and the relative intensity of an earthquake.

Palmieri was quite a character, beavering away taking measurements in the observatory – less than a mile from the crater – while at least five volcanic eruptions took place. During one of these eruptions, on 24 April 1872, Palmieri left the observatory to pick up an instrument in Naples, and managed to climb back up to it, barricading himself inside. He was surrounded by torrents of molten lava, and volcanic rocks crashed down all around him. He later wrote that the eruption was a "sensational show".

TELEGRAPH

MACHINE NUMBER 055

The telegraph worked by transmitting electronic signals along wires that connected two "stations", also known as telegraph offices. The telegraph completely revolutionised long-distance communication.

As for who invented it, it's one of those stories where two sets of teams were working on it at the same time, one team in the UK, and one in the US. The English inventors William Cooke and Charles Wheatstone came up with a device that featured a number of magnetic needles on a board displaying the letters of the alphabet and numbers. An electric current moved the needles to point to a different letter or number. They patented their telegraph system in May 1837 and it was already being used on the Great Western Railway between Paddington Station in London, UK, and West Drayton, 21km (13 miles) away, by the following year.

Meanwhile, Samuel Morse, a portrait painter turned academic, developed a fascination for electromagnetism in 1832, after a chance meeting with scientist Charles Thomas Jackson on a transatlantic crossing. In 1837, Morse developed a telegraph that involved pushing a key to complete an electrical circuit. He collaborated with a fellow academic at the University of New York, Alfred Vail, who developed an instrument to record the signals.

In 1838, Morse and Vail demonstrated their machine at Speedwell Ironworks in Morristown, New Jersey. They'd connected 3.2km (2 miles) of copper wire between the transmitter and receiver, and, in numerical code, Morse sent the message, "A patient waiter is no loser".

In the early 1840s, Morse and Vail developed what became known as Morse code – a system of dots and dashes to represent letters and numbers. Similar to the typewriter's invention (see page 80), they gave simpler, shorter codes for the letters and numbers that were used most often. In 1844, Morse and Vail sent a telegraph message from a transmitter in the Supreme Court at the US Capitol in Washington D.C., to Mount Clare Station in Baltimore, Maryland, over 65km (40 miles) away. It read, "What hath God wrought?".

The telegraph system took off after that, but the next major challenge was how to lay a telegraph cable under the sea to connect two landmasses. The first one linked St Margaret's Bay in England and Sangatte in France, a distance of approximately 40km

IT WAS AN UNBELIEVABLE TRIUMPH OF ENGINEERING AND INVENTION EVEN IF IT TOOK 67 MINUTES TO SEND THE MESSAGE (TWO MINUTES FOR EACH CHARACTER).

(25 miles), and was completed in 1851. Just five years later, the American businessman Cyrus Field formed a company with the plan to lay a telegraph cable thousands of metres below the surface of the Atlantic Ocean. The idea was to use two former warships – the USS *Niagara* and the HMS *Agamemnon*, to lay the 4,000km (2,500 miles) of cable. Incredibly, in August 1858, the venture succeeded, linking Valentia Island on the west coast of Ireland with Newfoundland, off the North American mainland. The first official message was from Queen Victoria to US President James Buchanan on 16 August 1858. It was an unbelievable triumph of engineering and invention, even if it took 67 minutes to send the message (two minutes for each character) and another 16 hours to receive confirmation that it had worked. Unfortunately, the cable broke a few weeks later and it wouldn't be until 1866 that another cable was successfully laid from the ship SS *Great Eastern*, constructed by Isambard Kingdom Brunel and John Scott Russell. This time the cable worked dependably and could transmit eight words per minute, forever changing the communications network.

TELEGRAPH
The invention of a machine that could transmit and receive electronic signals along wires completely changed the face of communications.

WRISTWATCH

According to Guinness World Records, the first wristwatch was made by the legendary Swiss watchmaker Patek Philippe in 1868 for Hungarian aristocrat Countess Koscowicz. She'd commissioned him to design and make it as a stunning piece of decorative jewellery, and it was certainly that. The watch itself is actually quite simple, a round face sitting within a gold square, but it's flanked by ornamental floral designs featuring rose-cut diamonds decorated with black enamel, contrasting with the gold.

But it's just one in a whole range of "first wristwatch" stories. Queen Elizabeth I was presented with a bracelet containing a spring-driven clock by her favourite, Robert Dudley, the Earl of Leicester, in 1571, although it was intended to be worn on the arm. The trouble is, the item no longer exists and no one knows who made it, so we can't really be sure exactly what it looked like. Then you've got Blaise Pascal, the genius French mathematician, physicist and philosopher, who started wearing his pocket watch on his wrist in the middle of the 17th century, but that's just cheating. The other origin story that has really stood the test of time (sorry – I had to get one pun in there) and competes with the Philippe one, concerns a watch created by Breguet, another legendary watchmaker.

In 1810, Caroline Murat, the Queen consort of Naples and Napoleon Bonaparte's younger sister, commissioned Abraham-Louis Breguet to create a timepiece mounted on a bracelet. He finished it in 1812 and it really had all the bells and whistles – it was a repeating watch, so it chimed the hours; it was equipped with a thermometer and a moon-phase indicator; and was mounted on "a wristlet of hair and gold thread", according to Breguet's website.

In 1880, Kaiser Wilhelm I, the German Emperor, spotted the potential for the wristwatch to become more than a fashion statement, apparently after learning that an officer had complained that timing an artillery bombardment with a pocket watch was proving tricky. So the Kaiser commissioned Swiss watchmaker

BLAISE PASCAL, THE GENIUS FRENCH MATHEMATICIAN, PHYSICIST AND PHILOSOPHER, STARTED WEARING HIS POCKET WATCH ON HIS WRIST.

Girard-Perregaux to make 2,000 wristwatches for officers in the German Imperial Navy, and brought over the watchmakers from Switzerland to Berlin to complete their task. This might have been the first wristwatch for men, and it was certainly the first mass-produced watch. Brits were thinking along similar lines to the Germans and some officers were photographed wearing wristwatches during the Third Burma War of 1885–1887. In the early 20th century, watches were specialised for military use and by 1916, in the middle of the First World War, "luminous wrist watch with unbreakable glass" appeared in the "Officer's kit for the front" (in the book *Knowledge for War: Every Officer's Handbook for the Front*); its importance is emphasised by the fact it appears first in the list, above both "Revolver" and "Field glasses". Some watches were even designed with "shrapnel guards", a pierced metal sheath that slipped over the watch so the owner could see the time through the holes.

After the First World War, the wristwatch became an essential item, with dozens of watch manufacturers setting up shop. This flurry of activity helped lead to the self-winding mechanism, developed by John Harwood in 1923. Until then, watches were manually wound up by the wearer, but Harwood used the kinetic energy of the user's wrist motions to tighten the mainspring. Sadly, his business didn't survive the Great Depression following the Wall Street Crash of 1929.

WRISTWATCH
First developed as jewellery for women, watches became an essential item for men after the First World War.

DENTAL DRILL

MACHINE NUMBER 057

Ah, the dreaded dental drill. Astonishingly, dentistry was being practised by the Indus Valley Civilization in 7000 BCE, in what is now Pakistan. Teeth estimated to be over 9,000 years old, have been found, with concentric grooves consistent with some form of a specialised bow drill used to hollow out tooth cavities.

Just to put this date into perspective, earthenware – the first kind of pottery – only dates back around 9,000 years, and rice was only cultivated from around 5700 BCE. That gives us a window into how much our ancestors valued sorting out dental problems.

We have to jump forward to the 19th century to the next major development, and that was the drill invented by the British dentist George Fellows Harrington in 1864. It was called *Erado,* which is Latin for "I scrape out", and was powered by a clockwork mechanism – operated by a key – that was wound up and then released, which gave the dentist two minutes of drilling time. It was the fastest dental drill in existence at the time, and Harrington came up with all sorts of developments, including interchangeable heads and cleverly angled handpieces for those hard-to-reach areas. It wasn't a huge success, though – it was hard to control, not to mention noisy, and was overtaken by a more efficient machine a few years later. Harrington's Erado is in the Science Museum in London and it's incredibly beautiful, with its silver cylinder engraved with intricate floral decorations and velvet-lined walnut case. Everything was so well considered. It actually looks more like a fine watch than a drill.

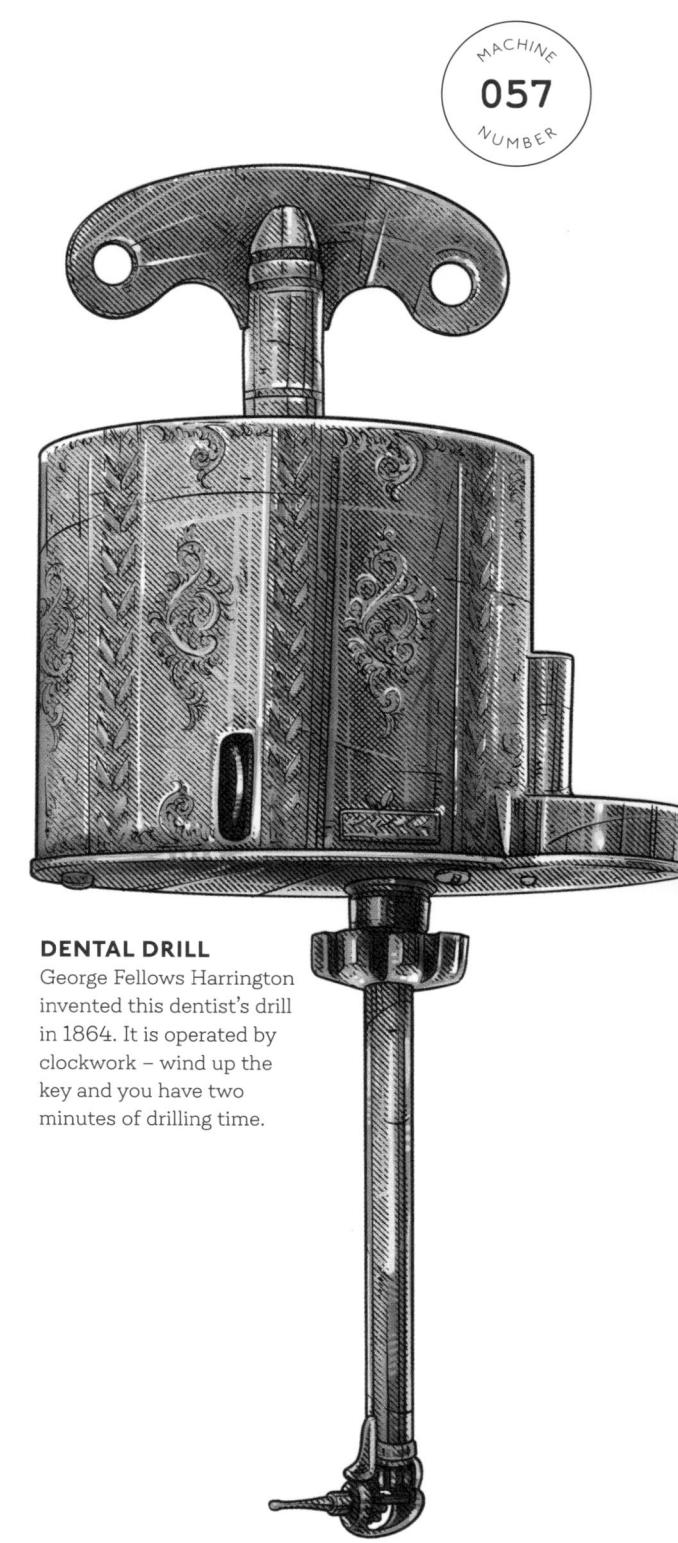

DENTAL DRILL
George Fellows Harrington invented this dentist's drill in 1864. It is operated by clockwork – wind up the key and you have two minutes of drilling time.

In 1872, the American dentist James B. Morrison adapted the foot treadle from a sewing machine to develop a foot-operated dental drill, which could reach 2,000rpm, but this had the slight drawback of needing either the dentist, or an assistant, to keep working away at the treadle. Everything changed when the first electric drill was designed by the American dentist George F. Green. Suddenly, there was a drill that could maintain drill-speed and didn't need two people to operate it. In case you're wondering about the pain relief options at the time, nitrous oxide (laughing gas) had been used in dentistry since 1844, thanks to the experiments of the American dentist Horace Wells. He famously had his assistant, John Riggs, extract one of Wells's teeth while under nitrous oxide. Talk about putting your money where your mouth is.

I've got various dentist's tools in my workshop because they're small, precise, durable, very well made – and perfect for delicate metalwork. I was using one recently to remove a clutch plate from an old Triumph motorbike. Some crafts are so well established that they have very specific small tools, like horology and lutherie (stringed instrument-making), but the rest of us have to be resourceful, think outside the box and adapt. Craftspeople like Lucia Scalisi (our painting conservator) and Kirsten Ramsay (our ceramics conservator) from *The Repair Shop* have all sorts of dentist's tools in their drawers.

I had a dentist's appointment soon after researching dentist's drills for this book, and I couldn't help myself asking her all sorts of questions about what type of drill bits she used, whether they were high-speed steel drill bits, or carbon. And she loved being asked (I'm guessing it doesn't happen that much) and told me all about them. As with many crafts, there are very specialist manufacturers who produce these super-precise machines for every dental operation. They actually use very similar rotary tools to the pneumatic die grinders that I use in metalworking.

I'VE GOT VARIOUS DENTIST'S TOOLS IN MY WORKSHOP BECAUSE THEY'RE SMALL, PRECISE, DURABLE, VERY WELL MADE – AND PERFECT FOR DELICATE METALWORK.

STEAM ENGINE

MACHINE NUMBER 058

Steam engines go back way further than you might think, to the 1st century CE and the legendary Ancient Greek inventor Hero of Alexandria.

Hero called it the Aeolipile and basically thought of it as a toy. The design involved a hollow sphere with two curved nozzles projecting outwards in different directions. The sphere was mounted between two upside-down L-shaped supporting arms rising up out of a covered cauldron beneath. The supporting arms look structural at first glance, suspending the sphere above the cauldron, but one of them is actually a steam pipe. When the cauldron is heated, high-pressure steam shoots out of the nozzles, making the sphere rotate.

In 1698, the English inventor and engineer Thomas Savery came up with a steam-powered pump to raise water. He was granted a patent, but it was one of those amazingly broad patents which basically covered any pumping device that raised water using a heat source. The upshot of this was that the engineer and inventor Thomas Newcomen, who'd been working on a more advanced steam engine (technically known as a beam engine) to pump water, around 1712, had to come to an arrangement with Savery. Newcomen's invention involved an open-topped cylinder that housed a piston, connected by a chain to a wooden rocking beam, with another chain holding a counterweight on the other side of the beam. A furnace underneath sent steam into the cylinder, drawing the piston upwards. Then cold water was released via a valve into the cylinder, condensing the steam (turning the steam back to water) and creating a vacuum. It's this vacuum that pulled the piston back down to the bottom of the cylinder. That action closed a valve between the cylinder and the piston, allowing steam to fill the cylinder again and send the piston back upwards. This engine was used successfully for nearly 50 years to pump water out of mines – a serious improvement from the expensive and agonisingly slow technique involving hundreds of horses hauling buckets on rope. Scottish engineer James Watt was repairing a Newcomen engine in 1764, and couldn't help becoming frustrated with the amount of steam it wasted. So he came up with a major improvement to help efficiency, by adding a condensation chamber connected to the cylinder. This meant that the cylinder wasn't losing so much heat when steam condensed into water. He patented his invention in 1769 and, after partnering up with entrepreneur Matthew Boulton in 1775, Watt had the funding and facilities to produce a test engine. When it was completed, it used around 75 per cent less fuel than Newcomen's engine. One of Watt's most famous beam engines, which became known as "Old Bess" was manufactured by Boulton & Watt in 1777 and was still in operation until 1848. It's now on the ground floor of the Science Museum in London, next to the gift shop.

Watt next set his sights on adapting his machine to provide rotary power. And with rotary power, you can drive machinery, and that was a complete game-changer. He achieved this with an arrangement of cogs known as the sun-and-planet gear, which converted the up-and-down motion of the beam to rotary motion through two cogs. One cog was attached to the end of a rod connected to the end of the beam; this cog moved in an orbit (hence the sun-and-planet name) around the teeth of the second cog, which was mounted on a flywheel attached to the drive shaft. Watt also developed the double-acting piston – a piston that could "drive" on both its forward and backward strokes. This made the whole operation faster, more powerful and more efficient.

There are still working steam engines, preserved by dedicated volunteers, including the 1812 Boulton & Watt engine at Crofton Pumping Station in Wiltshire, UK, which runs on weekends in the summer. And it still does the job it was designed to do. You can find incredible old steam engines at rallies and fairs all over the country. The Weald & Downland Living Museum has a steam fair in August every year, and engines arrive from all over the country in the week beforehand, thumping away as they get closer and closer, the wheels crunching on the gravel. We have to stop filming because they come in on the track behind the barn and make so much noise. But it's a perfect excuse to sneak out the back, have a chat with the owners and try and get a ride on one of them. They're epic machines.

I know I love my cars, vans and motorbikes, but one day I'm going to restore a steam engine. It's on my bucket list. I'm not obsessed yet, but I love and admire them. The engineering is just incredible, the coachbuilding is immaculate and the signwriting is stunning. Plus you have to make absolutely everything because there's nowhere to buy the parts that you need, so it's a daunting but exciting challenge. And that's what I'm in the market for.

STEAM ENGINE
The invention of the steam engine made pumping water out of mines much, much easier.

STEAM LOCOMOTIVE

George Stephenson is the name everyone associates with steam locomotives, along with the famous *Rocket* of 1829, but there's much more to the story than that. Mainly that Stephenson wasn't the first to design or build a steam locomotive, he was just the first to build one that carried passengers on a public railway.

It was Cornish mining-engineer Richard Trevithick who constructed the first steam railway-locomotive. He was also the first person to successfully harness high-pressure steam in a steam engine, something that inventor James Watt thought was too dangerous to achieve. To be fair to Watt, boiler technology had moved on rapidly in the 1780s and 1790s since he had been at work. Using high-pressure steam (as opposed to low-pressure steam, which drove Watt's early beam engines) meant you didn't need a separate condenser. It also meant the engine could be smaller and lighter. Trevithick constructed his first steam locomotive, which he called *Puffing Devil,* in 1801. Driven by his cousin, Andrew Vivian, it took six passengers along a road in Camborne, Cornwall, UK, up a hill and along to the next village. Unfortunately, it broke down three days later during another test. Vivian and the other operators left the locomotive and snuck off to a local pub. Meanwhile, the water in the engine boiled away and the engine was completely destroyed. Trevithick didn't seem to see it as a setback, though – more a "user error" – and he got back to work on a new engine. In 1802, he was granted a patent for high-pressure engines, for both stationary and locomotive use, and in 1803, he designed a new engine, the London Steam Carriage, which Vivian drove through central London. That must have been quite a sight – and sound.

Later in 1803, there was another case of "user error", but this one had tragic consequences. The boy operating one of Trevithick's high-pressure stationary pumping engines in Greenwich, southeast London, forgot to release a valve before taking a break to catch eels in the River Thames. The resulting explosion killed four people. But it led to Trevithick adding a more sophisticated adjustable valve to the engine, and a safety valve sealed inside the boiler, which released when the pressure rose too high.

Also in 1803, Trevithick adapted one of his high-pressure steam engines that was driving a hammer at Pen-y-Darren Ironworks in south Wales, by putting it on wheels to make a locomotive. This locomotive became the subject of a famous bet between Samuel Homfray, the owner of the Pen-y-Darren Ironworks, and fellow iron-magnate Richard Crawshay. Homfray bet 500 guineas (the equivalent of well over £50,000 today) that Trevithick's locomotive could pull 10 tonnes (10 tons) of iron nearly 16km (10 miles) along a newly laid down railway. With some fanfare, the locomotive pulled the iron, along with five wagons and seventy men, the whole way in a little over four hours.

I've been on the replica of Trevithick's *Puffing Devil* down in Cornwall. I was on my way back from Penzance visiting my friend Julyan, the guitar doctor on *The Repair Shop,* and I saw a sign for a vintage rally organised by the St Buryan Agricultural Preservation Society. It was chucking it down with rain, but how could you not stop at that? It turned out to be the St Buryan Rally, not far from Land's End, and it had everything: steam engines, stationary engines, locomotives, classic tractors, classic cars – so many Series 1 Land Rovers – and even a steam-powered racing car. It was a rebuild of the famous *Whistling Billy*, which had been destroyed in 1912. But the replica of the *Puffing Devil* – the first high-pressured steam locomotive to run on the road – really had pride of place there, barely 30km (20 miles) from Camborne, where it all started.

George Stephenson grew up in Wylam, on the banks of the River Tyne in northeast England. He started work at just eight years old to support his family, earning tuppence a day to keep an eye on a neighbour's herd of cows to make sure they didn't wander into neighbouring farms. "Geordie", as he was nicknamed, developed a fascination with engines and made models of them out of mud and clay. When he was old enough, he started work in the local coalmine, becoming an engine-man in his teens. But Stephenson couldn't read or write, so he went to night school at 18, realising how important it was to help him rise through the ranks. He got married in 1802, and they had their first child in 1803, Robert Stephenson. The couple moved into a house called Dial Cottage in 1804, 6.5km (4 miles) east of Newcastle city centre and close to Killingworth Pit, where Stephenson worked as a winch operator. Things were looking up when tragedy struck. Their second child died soon after birth. Ten months later, his wife died of tuberculosis. Stephenson wanted to move away from the area but returned after his father, also a pit man, was blinded in a mining accident.

Then he finally caught a break, in 1811. The pumping engine at Killingworth was knackered, but Stephenson reckoned he could fix it. And he did, superbly. So the bosses at the pit put him in charge of looking after all their colliery engines.

In 1806, Stephenson witnessed an explosion that killed ten of his colleagues. It was the age-old problem – miners needed to see what they were doing, but the only means of illumination was a lamp with an open flame. Not ideal when you're constantly surrounded by inflammable gas. Then there was the notorious mining disaster in 1812 at the nearby Felling colliery, which killed 92 miners. Things had to change, and Stephenson was one of the folks who was going to change them. So he developed a safety lamp – the Geordie lamp – that let air in via a perforated plate. If the air filled with flammable gas, the flame would go out, warning the miners to get out of there. The trouble was, the renowned scientist Sir Humphrey Davy was also working on a safety lamp. Both lamps had plusses and minuses. But it seems that it wasn't just about whose design was better; it was also the matter of a self-made pit-worker, with a thick northern accent, going up against a qualified, educated, knight of the realm. Davy's followers accused Stephenson of stealing his design. Stephenson was later exonerated, but it left a mark, probably motivating his decision to send his son Robert to private school.

In 1814, Stephenson designed a locomotive to haul coal and named it *Blücher*, after the Prussian general who rode to the Duke of Wellington's aid at the Battle

IT WAS ALSO THE MATTER OF A SELF-MADE PIT-WORKER, WITH A THICK NORTHERN ACCENT, GOING UP AGAINST A QUALIFIED, EDUCATED, KNIGHT OF THE REALM.

of Waterloo. He built it in the workshop behind his cottage. It could haul eight wagons filled with 30 tonnes (30 tons) of coal and reach a speed of 6.5km/h (4mph). Stephenson started building railways, which were then in their very early stages. He began to use wrought-iron rails instead of the brittle cast-iron ones that were prone to breaking. In 1821, he was hired to build the Stockton and Darlington Railway, with his 18-year-old son Robert's help. Stephenson chose a gauge of 1,435mm (4ft 8½ in), which became the standard gauge all over the world. Robert Stephenson and Company was set up to produce locomotives. The first one was called *Active*, later renamed *Locomotion* and on 27 September 1825, it hauled its first train. It carried 11 coal wagons and the first purpose-built passenger car. In all, around 450 people took the 40-km (25-mile) journey from Darlington to Stockton, travelling at speeds of up to 39km/h (24mph).

The Liverpool and Manchester Railway (L&MR) became the world's first intercity railway when it opened in 1830. Although, the year before it opened, the directors of the L&MR weren't convinced that locomotives were the best idea. They favoured using stationary steam engines positioned at various points along the line, which would pull wagons using cables. The Stephensons disagreed, persuading the directors to arrange a competition to see whose design would work best. This became known as the Rainhill Trials – to see who could design and build the best locomotive weighing under 6 tonnes (6 tons). It was a test of speed and reliability, because the line was intended to carry passengers more than freight.

Rocket was largely Robert Stephenson's creation, built at Forth Street Works in Newcastle. It featured a multi-tube boiler design (almost every boiler to that point just featured one pipe surrounded by water), and a firebox (the chamber where the fuel was burnt) with a good air supply and a saddle-shaped water jacket, which meant the water was in contact with the hottest part of the boiler. *Rocket* also featured an inclined cylinder rather than a vertical one and that meant the piston could be connected directly to the wheel – a much more efficient design. It also had a single pair of driving wheels with a small pair of rear wheels, and was made to accommodate just one driver. What's more, it was properly run in before Rainhill, unlike the other entrants. I think the Stephensons knew it would win. It was a reliable and already proven engine. They just needed it to work on the day. And wow, did it do that! Not only was their yellow-and-black *Rocket* the only entrant to complete the 97km (60-mile) race; it also reached a top speed of 48km/h (30 mph). The Stephensons collected the £500 prize and were awarded the contract to build locomotives

for the line. They hadn't just blown the competition out of the water, though. Their design provided the blueprint for locomotives for the next 150 years. It was a momentous day in the history of science and engineering. *Rocket* is currently back in the northeast where it was created, taking pride of place at the railway museum Locomotion in County Durham. Redevelopment is carried out at its usual home, the National Railway Museum in York.

STEAM LOCOMOTIVE
The Stephensons' *Rocket*, built in 1829, provided the blueprint for locomotive design for the following 150 years.

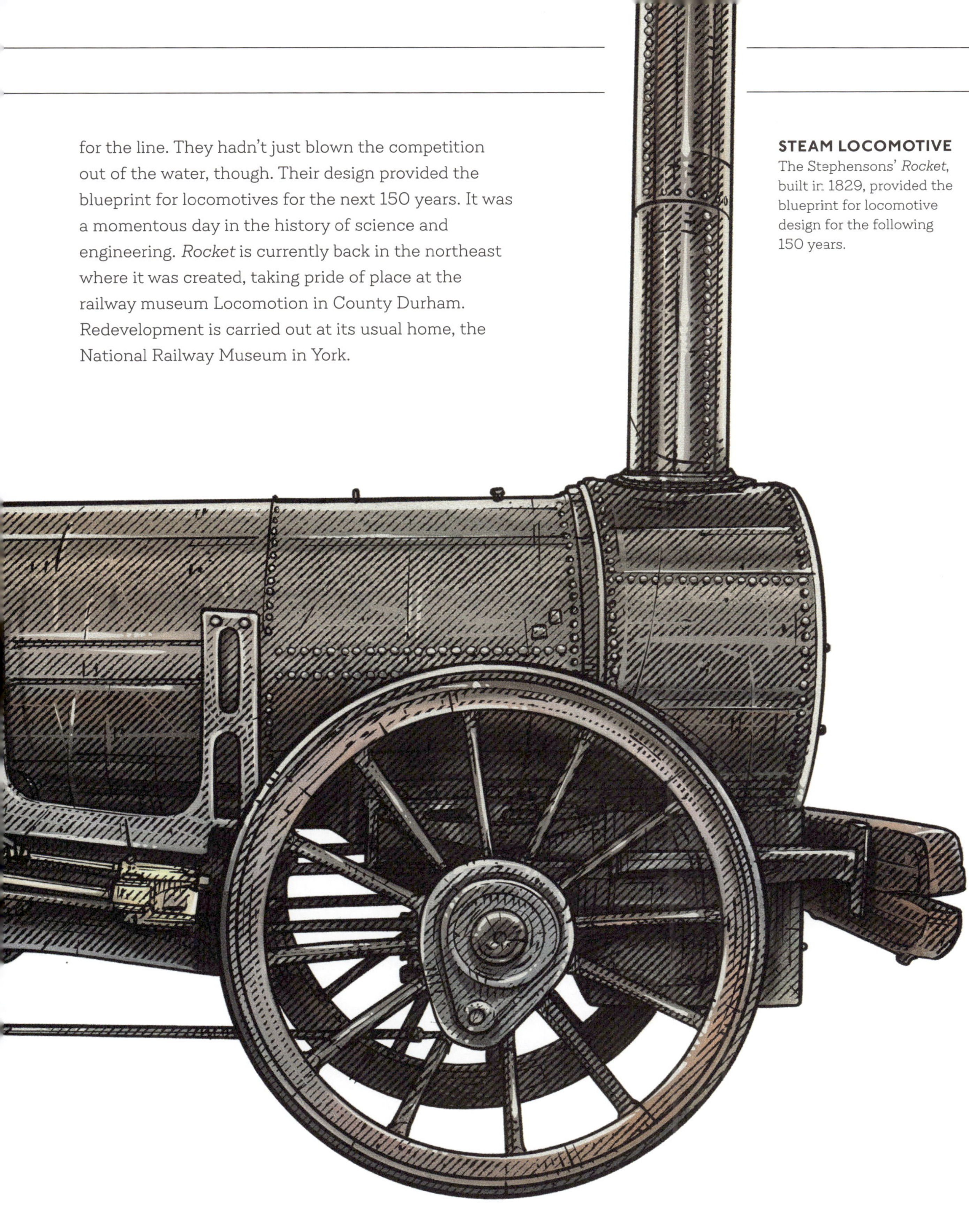

SCREW PROPELLER

MACHINE NUMBER 060

The idea of developing screws for propulsion has its roots with Archimedes' screw (see page 19), but it wasn't until the 18th century that a serious attempt was made to use the idea to drive ships through water. That came courtesy of David Bushnell, who designed a brass propeller for his submarine *Turtle* (see page 42), but after its sinking, he returned to army service.

The American inventor John Stevens constructed a 7.5m (25ft) boat, powered by a rotary steam engine, connected to a propeller with four blades, in 1802, but working with high-pressure steam engines was a pretty lethal business back then, and he opted to design boats with paddle wheels instead.

In 1835, two London-based engineers started working independently on screw propulsion – John Ericsson, a Swedish inventor and engineer; and Francis Pettit Smith, an English farmer turned engineer. Smith filed his patent first, got himself some funding, and built a 9m (30ft) canal boat to test his screw propeller (of two full turns) on Paddington Canal, to the west of London. Unfortunately, it was badly damaged in early 1837, changing it from a screw propeller of two full turns to a screw propeller of just one turn. Amazingly, this "modification" doubled the boat's speed to around 13km/h (8mph) and so Smith changed his original patent to include the upshot of his propeller's happy accident.

Later in 1837, Smith found himself frustrated by the Royal Navy's reluctance to embrace screw propulsion. "I'll show you lot," he probably said to himself before launching his revamped canal boat, this time fitted with a new single-turn propeller made of iron, along the Kent coast in full view of Royal Navy officers. It made impressive progress through choppy seas, and the Admiralty now approached Smith with their tails between their legs and invited him to spend some more time working on his method of propulsion. Smith, and several interested investors, formed a company and built a steamship together in 1838, fittingly called the SS *Archimedes*. It was the first steamship in the world to be driven by a screw propeller. And it proved to be massively influential on legendary engineer Isambard Kingdom Brunel, who was at the time working on the design for what was to become the largest passenger-ship in the world – the SS *Great Britain*. Brunel actually persuaded Smith to lend him the SS *Archimedes* to perform his own scientific tests on, and together they worked out the most efficient design for a screw propeller.

Ericsson had also experienced the Royal Navy's unwillingness to show much interest in screw propulsion in 1837, despite carrying out an impressive test on the River Thames in a 14m (45ft) steamboat in front of the assembled bigwigs of the British Admiralty. So Ericsson put two fingers up to the Brits and propelled himself westwards, finding a much more appreciative audience in America. He went on to design the USS *Princeton*, the US Navy's first screw-propelled steamer, which was launched in 1843.

SCREW PROPELLER
The Royal Navy was reluctant to embrace screw propulsion at first, but came round to the idea after a demo right under their noses left them in no doubt as to its potential.

SO ERICSSON PUT TWO FINGERS UP TO THE BRITS AND PROPELLED HIMSELF WESTWARDS, FINDING A MUCH MORE APPRECIATIVE AUDIENCE IN AMERICA.

BARBER'S CHAIR

The barber's chair has a proud American history. In 1878, the Archer Company of St Louis was the first to patent a barber's chair with a reclining back, and an attached footrest that lifted up and down.

They followed it up with an improved design where the chair could be mechanically raised and lowered. Eugene Berninghaus of Cincinnati, Ohio, designed the first barber's chair to recline and revolve, which he called the "Paragon". Berninghaus became a famous name in the barber's chair game, with some of his later models becoming props in iconic films, such as Charlie Chaplin's *The Great Dictator* (1940). But at the time, in the early 1880s, both Archer and Berninghaus were blown out of the water by Theodore Kochs.

Kochs was born in Germany in 1849. But by the time he was 18 years old, he had sailed to the US and found his way to Chicago. Two years later, he was running a pharmacy, but found his new life in tatters when the Great Chicago Fire of 1871 destroyed much of the city, including his livelihood. Like the chairs that he'd later invent, though, he was soon on the up again, founding the barber supply firm Theo A. Kochs Company, with an office and factory in Chicago's Fifth Avenue. And his barbers' chairs and barbers' furniture developed a reputation for their quality and durability. His trade catalogue of 1880 even went into detail explaining how his chairs worked: "Two rack-bars fastened on the upper part of the chair are connected with the gears, and the same gears are again in connection with a horizontal bar provided with a steel spring on the underside so as to form a Perfect Lock. The bar is operated by a lever with projecting pedal to the rear of the chair in easy reach for the barber's foot from all sides."

These chairs weren't just impressive functional items, though – they were also beautifully made, decorated and upholstered.

Mechanical barbers' chairs would change forever when Ernest Koken cranked things up a notch. The enterprising, 19-year-old Ernest had set up his company in St Louis, Missouri, to supply barbers with items such as shaving mugs, barbers' poles and barbers' coats; he also set up a side hustle, selling on used barbers' chairs. By the early 1880s, Ernest had a business partner, and they started making their own barbers' chairs. And in 1892, Ernest came up with the first hydraulically operated barber's chair. Koken's Congress Pedestal barber's chair is stunning, made of oak with an intricate fleur-de-lis detailing, mohair and leather upholstery, and an Art Nouveau-style cast-iron footrest. It's a work of art. After Ernest died in 1907, he left the company in the good hands of his son Walter, and his tenure – especially during the 1920s – saw one of the most successful periods for Koken. It even opened its own foundry in 1926, which could make 100 chairs a day.

It struggled, as so many companies did, during the Great Depression, and the 1930s was a real downturn for Koken. Trading picked up during the Second World War, with an increased demand for barbers' chairs, as well as tool chests, and cartridge cases that the company had been contracted to make. In the 1950s, it changed focus to produce more exclusive chairs for the posher barbers' shops that had appeared in airports and hotels. But its fortunes changed in the

1960s, not least because everyone wanted to look like John, Paul, George and Ringo, which didn't tend to involve regular trips to the barber. Things got worse with the arrival of a comparatively cheap, but smart and stylish, barber's chair called the Belmont. It became a massive hit, and one of them sits proudly in the front of my workshop.

I bought my Belmont when I was working in set design. I could have rented one but I had a look on some auction sites and a beautiful one appeared for almost the same cost as it would have been to rent it. Well, I say that – I was looking for an excuse to buy one, if I'm honest. Its first proper home was my old workshop in London, and it's travelled with me to the new place. I just love it. It's really comfortable to sit in and a lot of the time I use it when I'm on the phone. I love the fact that it always puts a smile on people's faces when they come to the workshop. They always want to sit in it and move the lever up and down. And that seems to activate fond memories of getting your hair cut when you were young. Maybe it's that feeling of life being simpler back then, or that unique period of calm in the chair. A classic barber's chair isn't just a machine; it's a beautiful time machine.

For me, the look of the Belmont transports me to the 1950s hot-rod era. It was such an iconic time, with the cars, the architecture, the colours… I love everything about it. The Belmont has big square arms that you can put a drink on, and there's an ashtray in one of the arms. It's like a Cadillac of a barber's chair and it actually feels a lot like you're inside a Caddy when you're sitting in one. It's got a lovely soft seat, lots of chrome and a latticed cast-iron footrest that looks a lot like the grille of a car. It's unashamedly flashy and showy, and everything's bigger and brighter than it needs to be. There are older barbers' chairs, and more beautiful ones, but the Belmont marries form and function so well. It was built to last, and it succeeds in that endeavour because so many barbers still use them (including mine). And I think it says a lot about the barber if they've chosen to invest in classic chairs like the Belmont. It speaks to the pride they have in their work, and their workplace. In a way, it sculpts the whole shop. But what amazes me more than anything about it – because it's such an iconic 1950s American design and even has a name that sounds like a famous New York hotel or bar – is that it's actually Japanese.

I worked on a Belmont barber's chair on *The Repair Shop*, which was the first time I've been given a fix to do that could have been lifted straight out of my workshop. It was a great feeling, because not only was I working on something I knew and loved, but it also equipped me with the skills to fix my one if anything ever goes wrong with it. The lady who brought the chair in was a barber, and came from a family of barbers. Her chair, which she'd inherited, had been broken when it was moved into her new barber's shop. It must have been heartbreaking, seeing something that had been in the family for decades in two pieces just before she was going to open her own shop. That's the time when you need every bit of help that you can get and luck to be on your side.

THE BELMONT HAS BIG SQUARE ARMS THAT YOU CAN PUT A DRINK ON, AND THERE'S AN ASHTRAY IN ONE OF THE ARMS. IT'S LIKE A CADILLAC OF A BARBER'S CHAIR.

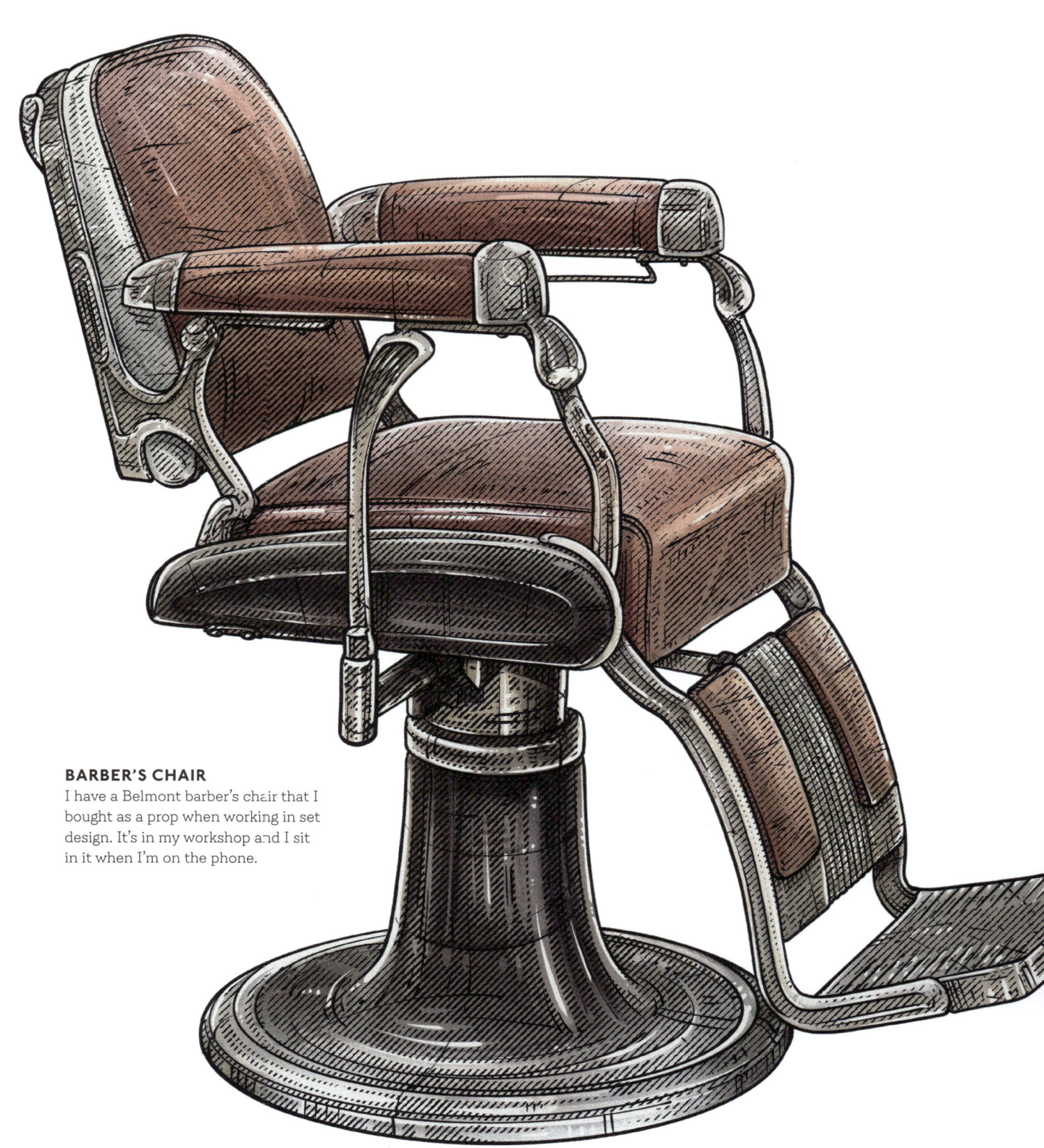

BARBER'S CHAIR
I have a Belmont barber's chair that I bought as a prop when working in set design. It's in my workshop and I sit in it when I'm on the phone.

Knowing that the chair was going to go straight back to use the day after we'd given it back to her, was a fantastic feeling. It doesn't actually happen that much on the show, but they're often my favourite fixes. It also cranked up the pressure knowing that there were hair appointments scheduled for just 24 hours later, so the chair had to be perfect.

Parts of the upholstery were salvageable, so we kept those and then went to great lengths to match the wear of the fabric. I remember getting in touch with a bunch of car-trimmer friends, calling in favours, even speaking to the parent company – the Takara Corporation – to get samples. In the end we caught a break with a car trimmer not far from *The Repair Shop* barn who worked on a lot of high-end cars. They had an archive of old bits of vinyl and details of where they'd got replicas from, and they found something that was nearly bang on. That was a terrific feeling, particularly because we'd had weeks of packages arriving and fabrics not being quite right. The colour didn't match, it was too soft, too rough, not thick enough.

I think you have a responsibility when you're working with such a classic, treasured object to do it justice. And then when the right one appears in front of you, sometimes in the most unexpected place… it's like you've found the pot of gold at the end of the rainbow.

CASH REGISTER

MACHINE NUMBER 062

```
The cash register was patented by James Jacob Ritty, a
saloon owner from Dayton, Ohio, in 1879. It was the third
model that he and his brother John, a talented mechanic,
had worked on, and the story goes that James came up
with it as a way to record each sale, so as to discourage
light-fingered bartenders from pocketing money. It wasn't
perfect – dishonest employees could still pocket takings
– but to be fair, they can still do that now on modern
registers, if they really want.
```

At first glance, Ritty's register looks like a keyboard with a large clock-like display above it, which showed customers how much they'd paid, and featured a locked compartment beneath for takings. He called it "Ritty's Incorruptible Cashier", which probably annoyed the honest bartenders in his saloon, and set up a company to manufacture and sell them. But after juggling both the saloon and the new business for two years, he sold his cash-register company and patents to a group of investors who set up the National Manufacturing Company (NMC). In 1884, the industrialist John H. Paterson (also a resident of Dayton, Ohio) became the majority owner of NMC. He'd bought three of Ritty's cash registers before then, and saw their potential. Paterson and his brother Frank renamed the company the National Cash Register Company (which makes sense) and it still exists today, although it was acquired by telecommunications giant

AT&T in 1991 and moved its headquarters in 2009 from Dayton to Duluth, Georgia.

Some of the cash registers that NCR developed, especially in the first two decades of the 20th century, were absolutely stunning and extremely well-engineered machines. One of its employees – Charles Kettering, who'd graduated with an engineering degree in 1904 and joined NCR soon afterwards – developed the first electrically operated registers, driven by an electric motor. Kettering later became a legend in the car world, inventing the first electric starter in 1911, which was fitted to the Model 30 Cadillac the following year. This development completely transformed the whole industry.

The introduction of NCR's Class 500 registers in 1908 was a landmark moment for the cash register – each model was electrically operated. In its 1909 trade catalogue, which you can find online, the No. 599 model had nine separate adding counters and cash drawers, so up to nine store clerks (or nine departments in a shop) could use the till at the same time. It was driven by an electric motor, and when you cranked the lever it performed over a dozen operations all at once, including: displaying the new indication, ringing the bell, unlocking the cash drawer and the clerk's individual cash drawer, adding the amount on the clerk's reel counter, and printing out the number of each sale, the initial of the clerk who made it, and the stock number of the item sold onto a detail-strip. And on top of the internal mechanisms, the Class 500 models featured beautifully intricate metalwork. The 599 model would set you back a princely $960 (the equivalent of over $30,000/£23,500 today). But what a machine! In 1912, NCR sold its one millionth cash register since 1889, and by 1915 it had 95% of the market-share of cash registers.

Steve, our horologist at *The Repair Shop*, fixed one made in the early 1910s, and I couldn't believe how intricate the detailing was or how complicated the internal mechanisms were. So many craftspeople must have been involved in making them, with their metal castings, carpentry and fine-art finishes. It was a serious investment, buying a cash register back then, and while they might have been conceived to deter theft, they became so much more than that: they became the centrepiece of a shop. The golden age of cash-register design ended after the First World War when necessity made them cheaper and more functional. But anything produced before that time, especially if you see the letters NCR, might well be a treasure waiting to be found.

THE NO. 599 MODEL HAD NINE SEPARATE ADDING COUNTERS AND CASH DRAWERS, SO UP TO NINE STORE CLERKS COULD USE THE TILL AT THE SAME TIME.

CASH REGISTER
Early electrically operated cash registers were a serious investment for a business, and a beautiful addition to a shop.

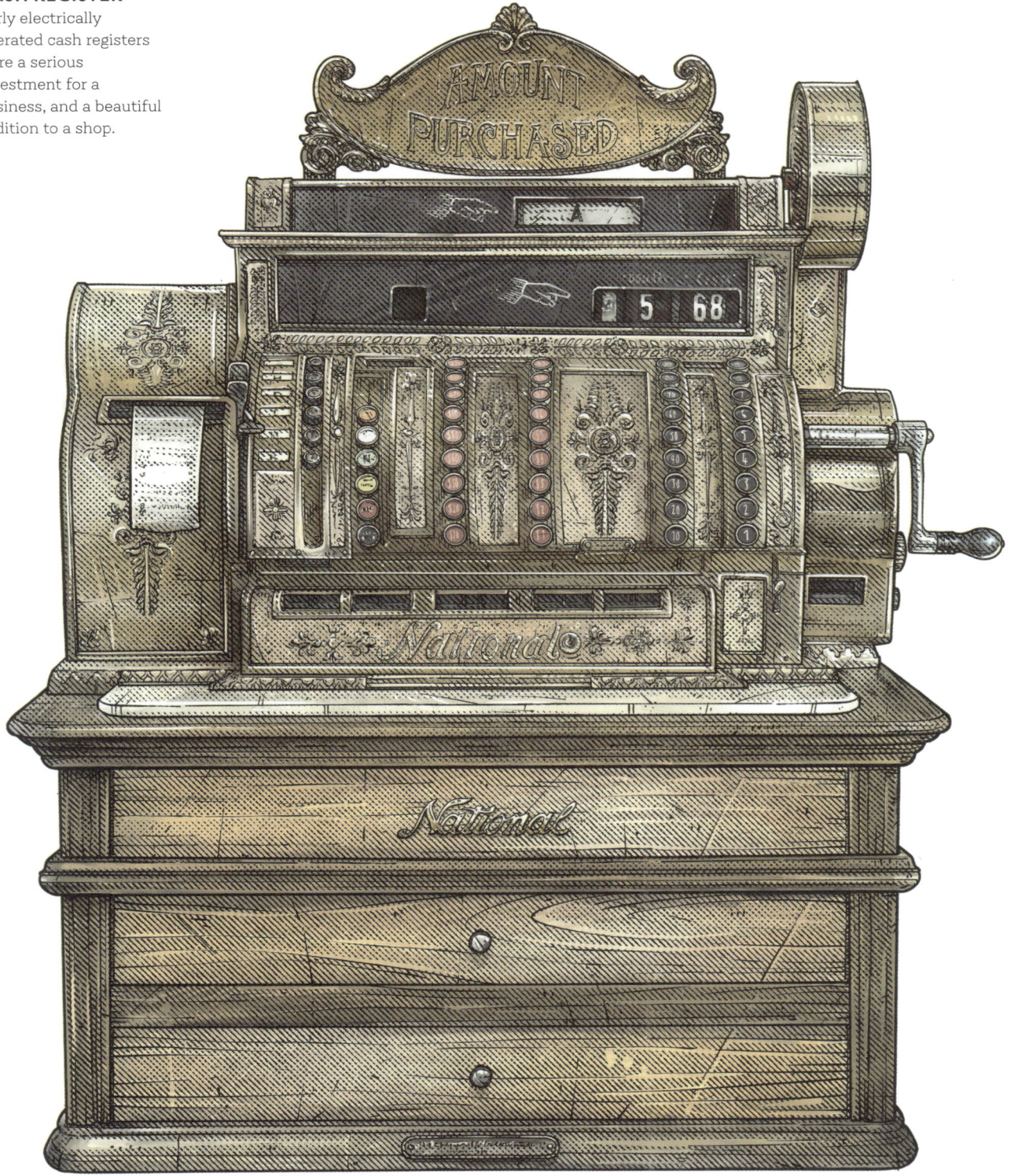

PATERNOSTER LIFT

MACHINE NUMBER 063

The Paternoster is a unique type of lift that is designed to be in constant motion. It consists of a chain of small, open-sided compartments that move in a continuous loop, travelling up one shaft, round at the top in a semicircle and then back down a second shaft, before again moving round in a semicircle, and up the other side. It can do this because the chain attached to the top of the compartments feeds into huge gears at the top and towards the bottom of the lift shafts. Faced with a paternoster lift, you'll see two elevator shafts, next to each other and no doors.

It earned its name from the chain, compartment and loop design, which looks like a set of rosary beads used to recite prayers. *Paternoster* is Latin for "Our Father", the first two words of the Lord's Prayer. Peter Ellis, a British inventor and architect installed the first paternoster Lift in 1868 in a famous office building he designed in Liverpool, called Oriel Chambers (named for the distinctive projecting bay windows, called oriels, that line the facades). The first paternosters were driven by steam engines, but electric motors became much more popular in the 20th century.

There are some big plusses of a paternoster lift over other lifts. Firstly, there's no pressing a button and waiting ages for a lift, before giving up the ghost and making for the stairs. There's also no stopping at each floor – you just step out at the right time, when the bottom of the compartment you're travelling in reaches the signposted floor. I know what you're thinking, though – isn't that really dangerous? Well, yes. It's not exactly wheelchair-accessible or child-friendly, and there have been fatal accidents, but those huge issues to one side, they are a lot of fun to ride. They did terrify some people when they were first introduced, though, because no one was sure exactly what would happen if the passengers stayed in a compartment after it went out of sight after the top or bottom floors. The urban legend was that they completely flattened. What actually happens is that they travel in a semicircle, right past the huge gear.

Concerns about safety meant that very few new ones were built after the 1970s, but they are very popular machines and people seek out surviving examples. There are a number that still work in the UK and Europe, including many in Germany, the Czech Republic and Slovakia. According to Guinness World Records, the largest operational paternoster lift is the 38-compartment paternoster lift in the 78m/256ft-high Arts Building at the University of Sheffield, England. I've never ridden in that one, but I have been in the one inside the Attenborough Tower at the University of Leicester in England. The first thing I did was just stand and watch it – it's an incredible machine. It hypnotises you a bit, with its creaking, clanking sound, and the constant, slightly sedate motion. It's such an efficient way to travel up and down, but I couldn't help but wonder about what happened when each compartment reached the top. Now I know you're not supposed to, there were signs

THE FIRST THING I DID WAS JUST STAND AND WATCH IT, IT'S AN INCREDIBLE MACHINE. IT HYPNOTISES YOU A BIT.

up saying *don't do it* and I'm absolutely saying *don't do it*, but I trusted the mechanics enough to know that the compartment wouldn't suddenly flatten, as the urban legend had it. So I stayed on board and can confirm that the compartment travels to the top of the chain, rotates to the left and comes back down. It was installed in the 1960s, but the university had to close it in 2017 because it was too expensive to maintain. There was a petition to save it which attracted thousands of signatures, but was sadly unsuccessful. People really loved that lift, and I can completely understand why!

PATERNOSTER LIFT
The ingenious design of the paternoster lift meant you didn't have to wait too long for the lift to arrive – although there was a bit of a trip risk getting in and out.

ESCALATOR

MACHINE NUMBER 064

The moving staircase was another great American invention of the 19th century. As for who invented it, it's one of those machines where someone designed a machine but failed to build it. And that person was Nathan Ames, a patent solicitor living in Massachusetts, US, who invented the concept of the escalator and patented it in 1859.

He called it "Revolving Stairs" and it featured two different designs: the first design had the stairs arranged side by side and attached to an "endless" belt that revolved around two large wheels; the second design was a triangular arrangement with an ascending and descending flight of stairs, and a (presumably hidden) horizontal section of belt and stairs forming the "base" of the triangle. Ames intended that the wheels could be turned by hand-cranking them, via a system of weights, or powered by a steam engine, but none of that came to pass. It was the same story with another US inventor, Leamon Souder from Pennsylvania, who patented his "Stairway" (not the most imaginative name) in 1889 in Philadelphia, along with patents for three further escalator-like machines, but they weren't built either.

The man who did get the wheels turning was the engineer Jesse Wilford Reno, who invented the "Endless Conveyor or Elevator" and patented it in 1892. It was a conveyor belt that would move slowly at a 25-degree angle and riders would stand on parallel cast-iron cleats mounted on the belt. The whole thing was powered by a hidden electric motor. The first of these conveyors was installed in 1896, as a novelty on the Old Iron Pier, on Coney Island in New York, and it was a big hit in the two weeks it spent there. Although, if we're being technical, it didn't have steps, so we should count it as an inclined walkway rather than an escalator. Reno also designed a spiral moving-staircase designed to run in a continuously clockwise direction, and one was even installed as an experiment at Holloway Road underground station in London in 1906. It was intended to take passengers from the platforms to street level in around 45 seconds, but the design was apparently flawed and was probably never used. It was taken apart in 1911 and forgotten about, but a surviving part was discovered in 1988, and a 4m (13ft) section was restored in 2010. It's now in the London Transport Museum's Acton Depot.

The first "moving staircase" in the UK – and this is a good pub quiz question – was installed in November 1898 by French manufacturer Piat at the famous department store Harrods in South Kensington, London. Incredibly, nervous customers were offered smelling salts or a glass of brandy by employees after they reached the top. I wonder how many folks rode it multiple times claiming to be overcome with panic. It wasn't what you or I might think of as an escalator, though. It was more like an inclined conveyor belt made from woven leather, and featuring a mahogany and silver-plate glass railing.

The "escalator" was the name trademarked by the American inventor Charles Seeberger in May 1900 for a moving staircase exhibited at the Paris Universal Exposition, which was held between April and November 1900. The escalator was a collaboration between Seeberger and the Otis Elevator Company, which actually produced it, with the first one made at the Otis factory in Yonkers, New York, in 1899. Seeberger, who'd worked on his own moving staircase design from 1895, bought fellow inventor George Wheeler's 1892 patent to a moving staircase with steps that emerged and then flattened at the end (similar to the ones we know now, except this one required the rider to step off sideways at the end). Seeberger is thought to have come up with "escalator" from the Latin word *scala* meaning "stair" or "ladder". It was a big success, winning one of the Expo's Grand Prize awards, and soon major department stores and railway stations around the world followed suit.

ESCALATOR
First patented in 1859, the concept of a moving staircase didn't really become a thing until Charles Seeberger exhibited his escalator at the Paris Exposition in 1900.

BRAILLE-WRITING MACHINE

MACHINE NUMBER 065

The Perkins Brailler was a typewriter specifically designed to enable blind people to write using braille. It wasn't the first braille-writing machine — that was invented by Frank H. Hall, the superintendent of the Illinois Institution for the Education of the Blind in 1892. Hall's machine looked like a little piano, with six black keys that formed the letters, symbols and contractions used in Braille.

It was a revolutionary machine at the time for blind people, because, before then, blind typists used a slate-and-stylus writing-board to manually emboss each character. So it took ages, and it embossed Braille in reverse onto paper, so you couldn't see what you were writing. Hall's machine was simple and easy to use. Pins located on the underside of the paper (which you'd feed into the top of the machine) embossed the paper so it printed the correct way up. And you could write up to 50 words a minute. Hall never applied for a patent and he didn't sell his machine to make a profit. He was an inventor motivated by wanting to help people.

The Perkins Brailler was designed in 1951 by David Abraham, a teacher at the Perkins School for the Blind in Massachusetts. Abraham was born in northwest London in 1896. He became an apprentice to a gas-fitter after he'd left school, and joined the Army after the First World War broke out, aged 17, later signing up with the Royal Flying Corps (which became the RAF in 1918). After being discharged from the RAF in 1921, he returned to the UK and worked in the family staircase-making business, becoming a skilled woodworker and machinist. He was also an enterprising inventor, coming up with an idea for a staircase-moulding machine and applying for a patent for it. He emigrated to the US in the late 1920s with his family, before taking up a teaching position at the Perkins School for the Blind near Boston. In the summers, he'd work in the school's maintenance department. His talent in carpentry and with machines — he built a 7.6m (25ft) motorboat in a year, with the help of his class — didn't go unnoticed by the school's Director, Gabriel Farrell. Farrell asked Abraham to design a new type of lightweight, portable braille-writing machine to help the students at the school. As all craftspeople do (and I can certainly identify with David on this front), he spent hours of his spare time in his workshop building prototypes and testing out ideas. In 1941, he was happy with his prototype. But the Second World War put everything on hold, and it wasn't until 1951 that Abraham's machine was unveiled to the public. It was sturdy and didn't break down, and its light keys meant that even young children could use it. Over 375,000 of them have been sold in 170 countries around the world, making such a massive difference to blind and partially sighted typists. And Abraham's design and workmanship was of such high quality that the design basically remained unchanged until 2008 when a lighter, quieter version was released.

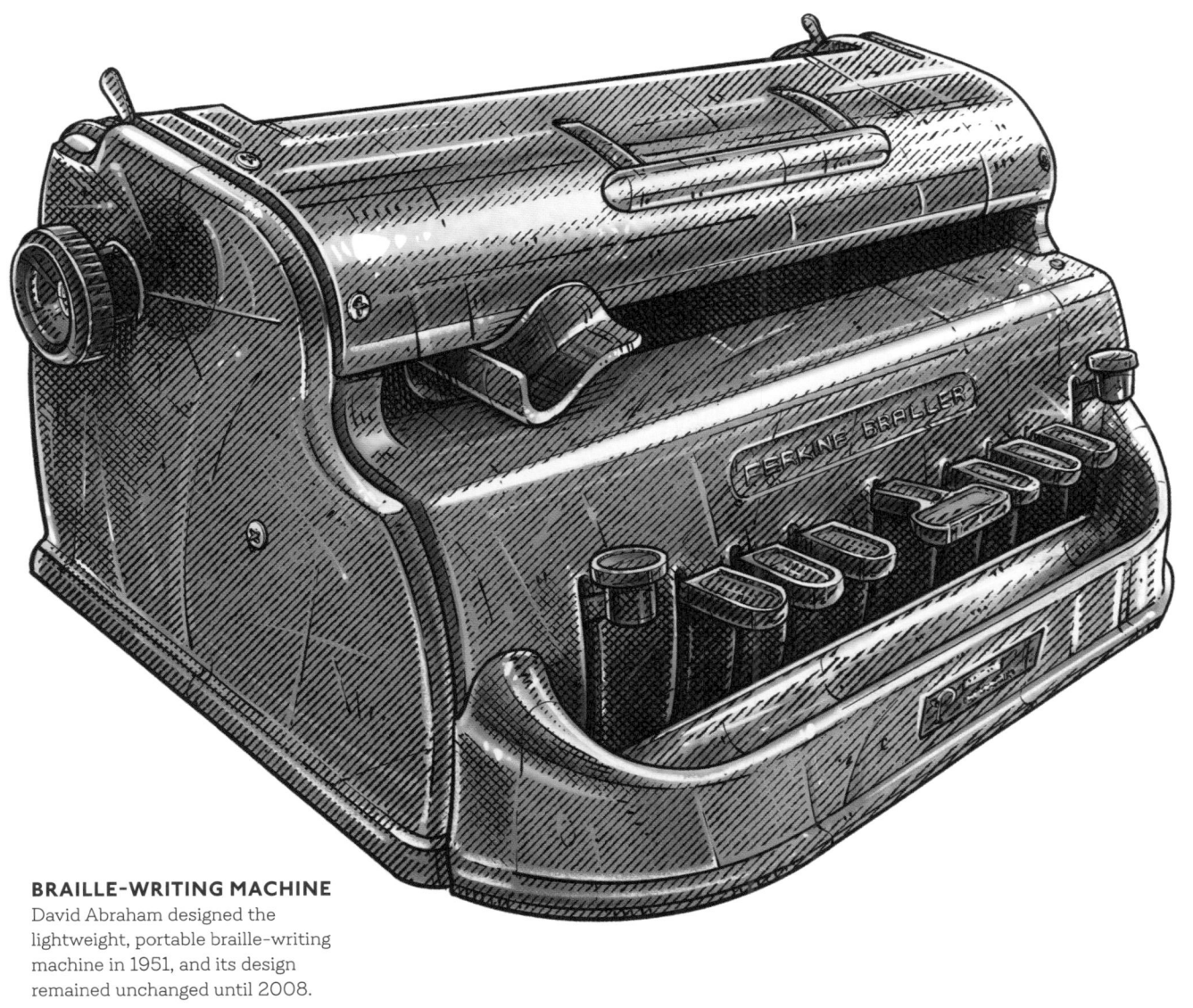

BRAILLE-WRITING MACHINE
David Abraham designed the lightweight, portable braille-writing machine in 1951, and its design remained unchanged until 2008.

HALL NEVER APPLIED FOR A PATENT AND HE DIDN'T SELL HIS MACHINE TO MAKE A PROFIT. HE WAS AN INVENTOR MOTIVATED BY WANTING TO HELP PEOPLE.

MILKING MACHINE

MACHINE NUMBER 066

In 1851, the first vacuum-milking machine was patented by Hodges and Brockenden, in England. It featured a suction cup that fitted over the udder and was connected to a hand pump.

Across the pond, American inventor Lee Colvin designed a hand-held pumping device in 1860, but it used teat cups rather than one cup that covered the entire udder. It caused quite a stir in the agricultural world, selling thousands in the US. Colvin even sold the patent rights to his machine in England for $5000, and over 1,500 of them were sold here. The problem was that exposing the teats to constant suction was damaging them and causing the cows some discomfort. It didn't help that sterilisation procedures were either non-existent or poorly understood.

The answer was the "Thistle", a pulsating milking-machine developed by Dr Alexander Shields of Glasgow, and this produced intermittent milk flow, mimicking the way a calf drank milk. It worked via a steam-driven pump, which combined suction and squeezing motions. It was approved by the US Dairy Association (USDA) and granted a patent in the US, but it proved to be tricky to clean and keep sanitised. But, it pioneered the use of pulsation, which is something that still exists in manual and automatic milking-systems today.

The next major milestone was the Surge Milker of 1922, which featured only four pieces of rubber to wash, short pulsation tubes, and a strong vacuum around the teats. It offered better milking, didn't damage the cows and avoided the issue of contaminated milk which happened if teat cups fell off and sucked up whatever was underneath the cow – which as you can imagine, isn't typically the most sanitary of environments. The Surge Milker arrived in Europe in the 1930s and the quality of the product, combined with strong advertising campaigns, meant it was a hit. By the 1950s, the one millionth Surge Milker had been produced. By 1955, it had over three-quarters of the entire US market share.

MECHANICAL PENCIL SHARPENER

Until the beginning of the 19th century, people used a small penknife to sharpen pencils, which, to be fair, can take a little while. The first mention of a *taille crayon* (French for "pencil sharpener") appeared in a book in 1822, stating it had been invented by C.A. Boucher, a French Army officer in the Engineer Corps.

But Boucher didn't patent it, and by all accounts it was complex and prone to breaking, and there's no evidence it was commercially produced. Just six years later, another Frenchman, a mathematician called Bernard Lassimone, patented his *taille crayon*, which basically comprised files embedded in a block of wood, and while he did bring it to market, it didn't become popular because, well, it sounds like it took about the same amount of time as whittling with a knife. The next development was in 1833 across the Channel with the invention of the Styloxynon by Robert Burton Cooper and George Frederick Eckstein, a kind of grinding box, which you used by moving your pencil back and forth in a V-shaped file. It wasn't a million miles away from Lassimone's invention, but it featured in all sorts of catchy ads in 1834, like: "The Stylixynon... gives the most delicate and durable point to Crayons, Black-lead or Slate-Pencil, with the utmost certainty, without requiring, like the Pen-Knife, any dexterity in using it – neither soiling the fingers with the scrapings of the lead or chalk."

In 1847, the precursor to the modern pencil sharpener, featuring a tube fitted with a tapering cone and blade that allowed every side of the pencil to be whittled at the same time, was invented by Thierry des Estivaux. Des Estivaux was an aristocratic French

Army officer, who'd fought at the Battle of Waterloo in 1815, aged only 17, alongside his father. He was also a budding inventor, having taken out a patent for a steam-powered propeller for coastal and inland navigation a year before inventing the sharpener.

Mechanical pencil sharpeners began to appear in the 1880s, and the US was the place to be. They all seemed to feature a lot of "P"s in their names, like the "Planetary Pencil Pointer", and the "Peerless Pencil Pointer", and most were made of cast-iron, featured a classic crank-handle with visible round disc mill cutters at the top, and were designed to be bolted to an office, or schoolroom, table or desk. The Olcott Climax Pencil Sharpener, sold from 1904, was one of the first with a cylindrical cutter containing 12 cutting blades arranged spirally. One turn of the handle would cut 56 very fine shavings off the pencil. It set the standard for the mechanical pencil sharpener for decades and wouldn't have looked out of place in a classroom in the 1980s.

The trouble with the mechanical sharpeners was that they weren't so great when you weren't at school or the office. People needed something portable, lightweight and user-friendly. In 1894, African-American carpenter and inventor John Lee Love came up with a clever development – a portable sharpener with a single blade fastened to a wooden casing that shaves the pencil down and catches the shavings until you want to empty them. It was patented in 1897, and you've probably got something very similar in your desk drawer.

I love mechanical pencil sharpeners, but I've got a confession to make. When I'm in my workshop, I'm often moving around and then suddenly need a pencil. But I never seem to have a pencil sharpener when I need one, so I tend to end up using a Stanley knife (which always seems to be right in front of me) as a makeshift sharpener, or sometimes the flat part of an angle grinder, or a bit of sandpaper. Whatever's around that is relatively sharp.

TICKET-PRINTING MACHINE AND TICKET-DATING PRESS

MACHINE NUMBER 068

Thomas Edmondson (1792–1850) was a trained cabinetmaker from Lancaster. He set up a furniture-making business with two friends in Carlisle, but it didn't work out and was declared bankrupt.

From then on, he worked a number of jobs, as an upholsterer, grocer and, later, a tea merchant. He caught a break in 1836, aged 44, when the Newcastle and Carlisle Railway was looking for staff. He became the Station Master at Milton railway station, just east of Carlisle. The ticketing system back then was slightly ridiculous, especially at smaller stations where the clerk in the ticket office would write the list of passengers (and their destinations) on a sheet of paper and hand it to the guard when the train pulled into the station. At larger stations, paper tickets were handwritten with the time of the train and the destination but still given to the guard rather than the passenger. Edmondson had a better idea. Thank goodness!

He came up with a design for a machine that would print railway tickets on card, each of which had a serial number, which meant the train company could count the number of tickets sold. When a ticket was issued, it was inserted into a dating press to stamp it and validate it. Using his woodworking skills, Edmondson made the printing block to print the tickets and the dating press out of wood, before designing an iron version which was made by a local clockmaker.

A specialised paper guillotine was designed to cut the card for the tickets into the correct size. And instead of the ticket office giving the passengers' tickets to the guard, they were issued to the actual passengers.

It was a complete game-changer.

Handwritten paper tickets became a thing of the past, as did the long queues at ticket offices. In 1840, he patented his machine, which could print 200 tickets a minute. The following year, he resigned and set up his own ticket-printing business. Thirty years later, his machines were printing 500 million tickets a year, in Britain alone. His machines took off around the world.

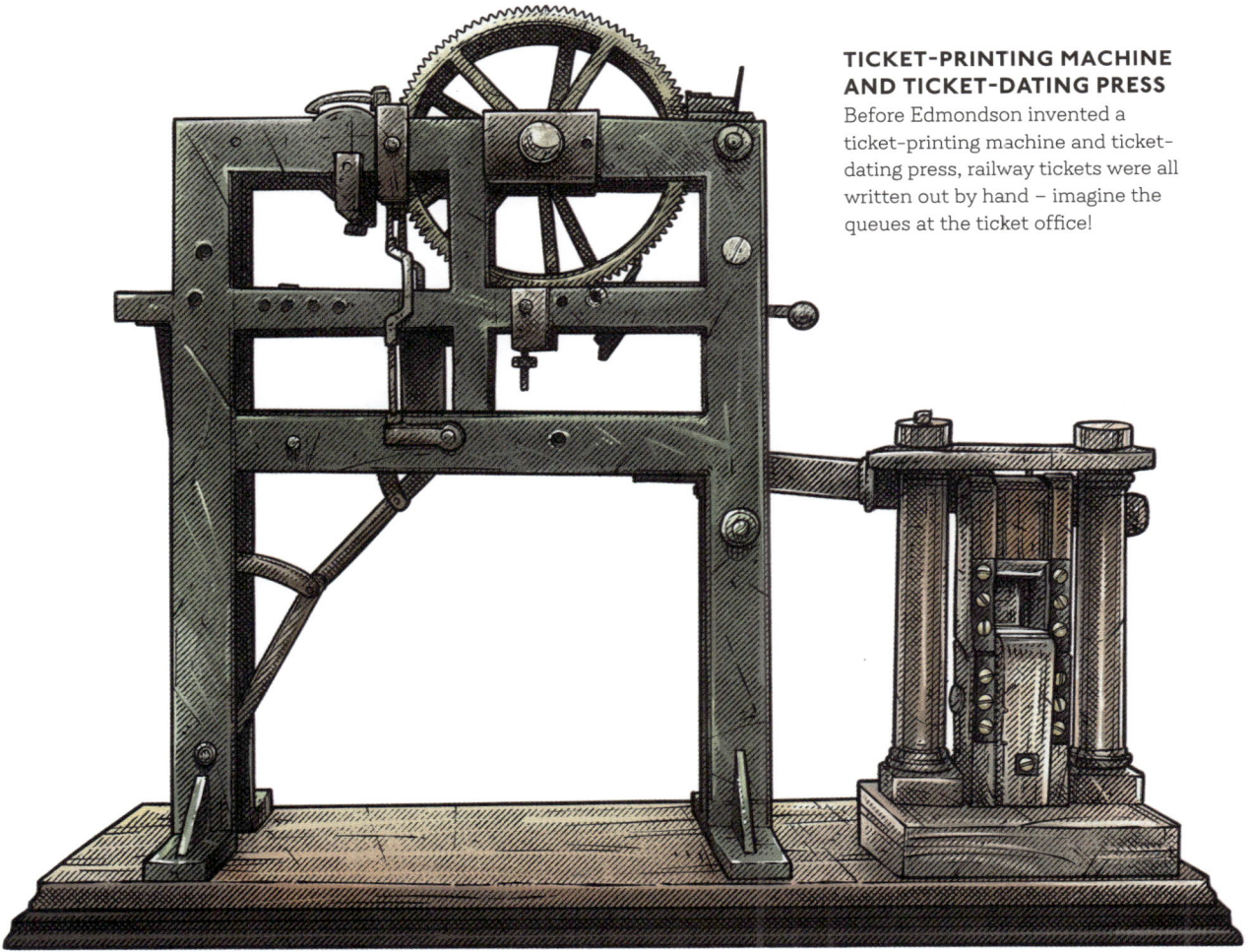

TICKET-PRINTING MACHINE AND TICKET-DATING PRESS
Before Edmondson invented a ticket-printing machine and ticket-dating press, railway tickets were all written out by hand – imagine the queues at the ticket office!

TIME RECORDER

MACHINE NUMBER 069

One of the first mechanical time clocks, used to record the time that employees arrived at and left work, was invented by New York State jeweller Willard LeGrand Bundy, around 1878.

The machine looked a lot like a pendulum clock with a keyhole below the actual pendulum, and each employee had a differently numbered key that they would insert into the hole. This motion turned a series of rollers inside the clock case which printed the time in hours and minutes, as well as the employee's key-number, using an ink ribbon on a roll of paper tape.

The following year, with his brother Harlow Bundy, they incorporated the Bundy Manufacturing Company, starting out with just eight employees. In 1890, they'd designed models for the UK market and had been awarded a patent. By 1898, the Bundy Manufacturing Company had made around 9,000 Bundy Time Recorders (which just became known as "Bundys"). The company merged with two other businesses to form the International Time Recording Company in 1900, based in Jersey City, New Jersey. This company was amalgamated along with three others in 1911 to form possibly the most boringly named corporation in history: The Computing-Tabulating-Recording Company (CTR). Unsurprisingly, the company's Canadian subsidiary chose a new name, and this one became familiar around the world: International Business Machines Corporation, or IBM for short. In 1924, the whole CTR business changed its name to match it.

The first card recorder, known as the Rochester Time Recorder, was invented by Daniel Cooper in 1894. Each worker had their own clocking-in card, and they inserted it into the machine and pulled a lever to stamp their arrival and departure times. This had a big advantage over a key recorder because you didn't need to produce a whole bunch of numbered keys for your employees. It didn't need a paper roll either, so the machine wasn't out of action for periods while you replaced the paper. This is another machine where the design actually changed very little in the best part of 100 years.

CAMBRIDGE ROLLER

MACHINE
070
NUMBER

The Cambridge roller is an agricultural roller invented by William Cambridge, an engineer, millwright and iron founder. He was active in Market Lavington, Wiltshire, England, in the 1830s–1840s, and made his name constructing portable steam engines (he was one of the first to do so), which were used to drive agricultural machinery. He also mounted the piston and cylinder over the engine boiler, which kept them warm and so improved the efficiency of the whole machine, saving money.

The Cambridge roller comprised equally spaced rings or ridges and was designed to break up clods and compact the earth to help it retain moisture and leave a firm bed for seeds. The ridges leave a series of grooves in the soil, which helps rainwater soak into the ground. There are various different modern variants of the Cambridge roller, with widely spaced rings, spirals, or "breakers", which are used to break up big clods. In the US, the Cambridge roller is known as the cultipacker. They are often used after sowing seeds as well, because they help ensure that seeds emerge at the same time.

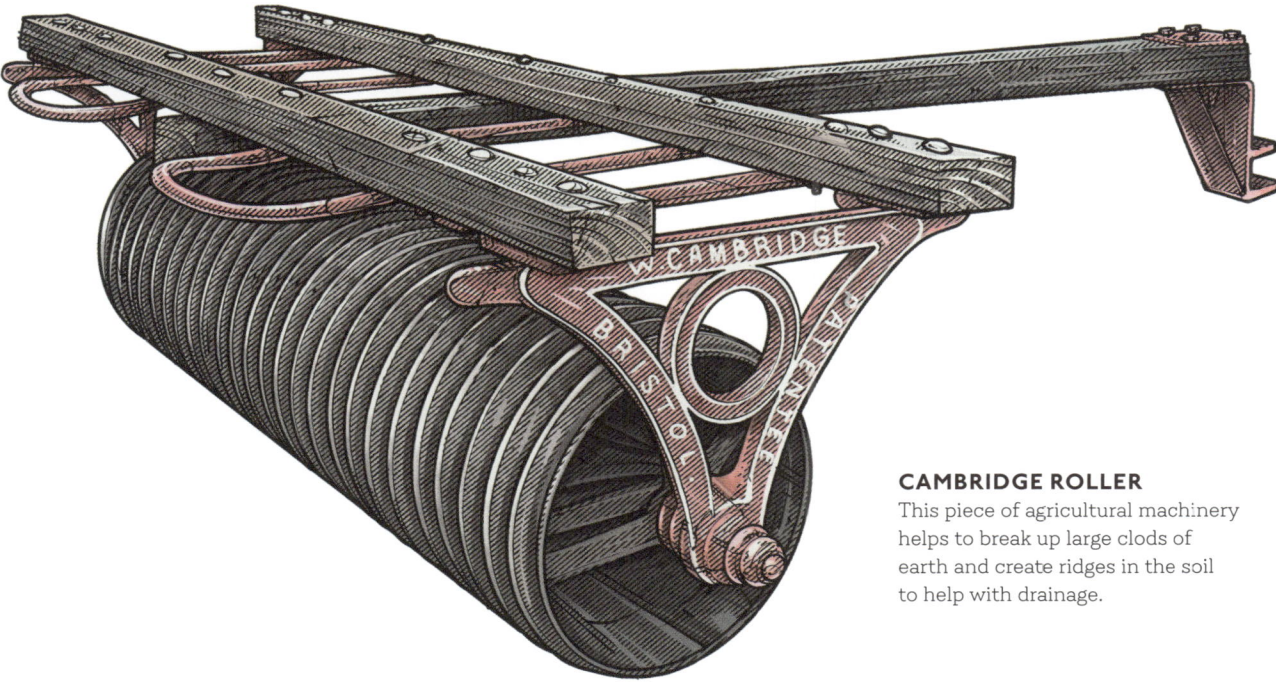

CAMBRIDGE ROLLER
This piece of agricultural machinery helps to break up large clods of earth and create ridges in the soil to help with drainage.

PUMPJACK

In 1859, Edwin "Colonel" Drake drilled the first commercial oil well in the US. It was in an area of northwestern Pennsylvania close to where oil had been found in surface pools. At that point in time, oil was being used to illuminate lamps, but it was also used for "medicinal" purposes to treat conditions such as rheumatism, burns and bruises, and even drunk as a laxative!

After experimenting unsuccessfully with other methods of extracting oil, he decided to drill down, like engineers would with a salt well, and hired an experienced salt-driller called Billy Smith. It wasn't understood that oil formed reservoirs in solid rock back then, so this was something of an educated gamble from Drake. So they bored into the ground with a drill powered by a steam engine, using sections of cast-iron pipe to stabilise the hole, until they hit bedrock, roughly 15m (50ft) down. It was slow-going from then on, but on 27 August 1859, Smith saw dark liquid. They'd struck oil, at 21.2m (69.5ft). They brought it to the surface using a basic hand pump and the oil was collected in a bath.

Although Drake's legacy was really the beginning of the petroleum industry in the US, he was more or less penniless by 1866, and his health was deteriorating. He finally caught a break when the state of Pennsylvania awarded him a life pension of $1,500 a year, in 1873, to recognise the fact that he'd made a lot of other people (and the state) very wealthy through his drilling techniques. However, another technique was needed to extract oil when the pressure in the well wasn't high enough to send the oil bubbling to the surface, something more powerful and sophisticated than a common-or-garden hand pump. The pumpjack was the answer.

The pumpjack was developed in 1925 by Walter Trout, who worked for the Lufkin Foundry and Machine Company, a manufacturing firm founded in Lufkin, Texas. The pumpjack features a long horizontal beam (the walking beam), which is anchored to the ground at its pivot point by two support beams, forming an A-frame. At one end of the walking beam is the "horse head" – the distinctive part of the pumpjack that seems to "nod" when the pumpjack is working. The nodding motion occurs because an internal combustion engine (or steam engine) turns gears that move a connected counterweight in a circular motion; the counterweight is attached to the walking beam via a vertical beam so the circular motion moves the walking beam up and down. At the end of the horse head is a sucker rod with a valve on the end of it (called the riding valve) and it's this rod (and valve) that travels downwards into the well. When it reaches the bottom of the well, the

riding valve closes and another valve (the standing valve) at the bottom of the well opens. This creates the pressure that allows oil to be drawn upwards. Each "nod" can draw up anything from 4 to 40 litres (1–10 gallons). And that distinctive motion and the machine itself is an iconic part of the American landscape.

PUMPJACK
An iconic part of the American landscape in oil-rich areas, pumpjacks are used to create enough pressure to draw oil up out of a well.

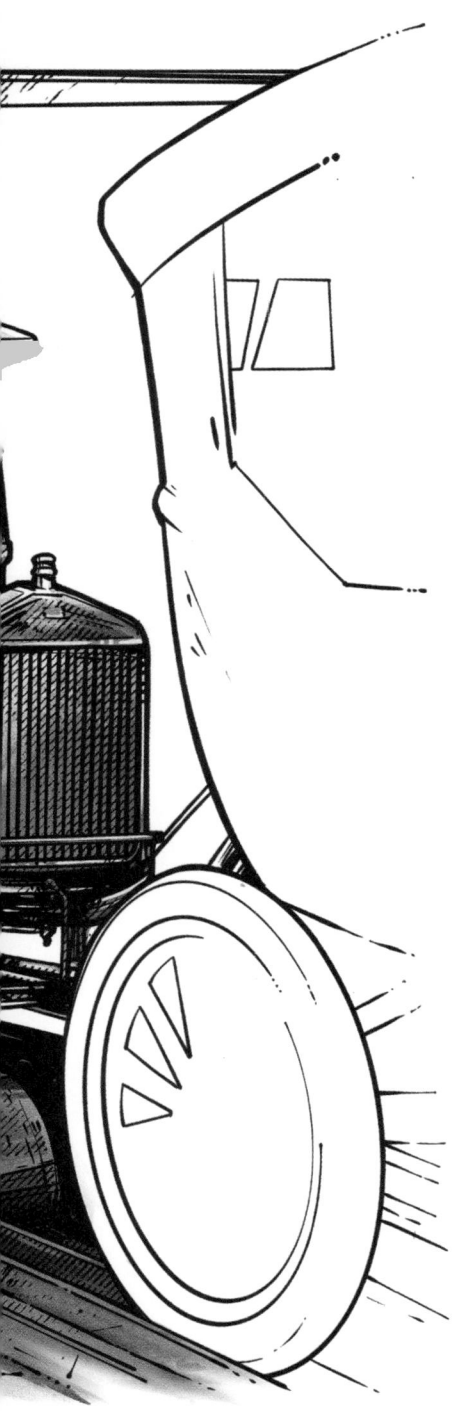

FORD'S MECHANICAL CONVEYOR BELT

MACHINE NUMBER 072

```
Henry Ford wasn't just an industrialist
- he was a trained machinist, engineer
and inventor who built his first motor
car in his spare time, while he was
working as an engineer for the Edison
Illuminating Company of Detroit.
```

In 1896, while still working at the company, he met his hero Thomas Edison, and they went on to become good friends. It was also the year that Ford built his ethanol-powered two-cylinder, four horsepower Quadricycle, which ran on four bicycle tyres, in a little workshop behind his home.

He left the Edison Company in 1899 and formed the Detroit Automobile Company, but it was shortlived, being dissolved in January 1901. Later that year, Ford built and raced a 26 hp motor car, founding the Henry Ford Company off the back of it, in October 1901.

In 1913, Henry Ford introduced the moving assembly line at his factory in Highland Park, Michigan. It completely transformed the way cars were made, with every aspect of vehicle assembly taking place in designated sections across the factory-floor using conveyor-belts inspired by the meatpacking plants in Chicago. Within six months of the introduction of the moving assembly line, the time it took to construct a Ford Model T (the first truly affordable car for middle-class Americans) had gone down from just under 10 hours to just under 6 hours, which is absolutely mind-blowing.

Although the wages were famously good at the Ford Motor Company – $5 for an eight-hour shift (along with a company profit-sharing plan) – the workers were subject to all sorts of controls. Firstly, the creation of the moving assembly line meant that they couldn't work at their own pace anymore. Also, their private lives were scrutinised by the company's Sociological Department, which made unannounced visits to employees' homes to check cleanliness, and monitored bank records and even the school-attendance of their children. Half their wage was granted only if they met "company standards for clean living". On the flip side, worker safety was prioritised at the factory and free English courses were offered to foreign employees, culminating with a diploma that counted towards their final citizenship exam to become American citizens.

FIDDLE DRILL

MACHINE NUMBER 073

```
Simon, a farmer from Halifax, West Yorkshire, brought
one of these on to The Repair Shop a few years ago. It
was one of the most obscure machines Suzie Fletcher and
I had ever seen on the show.
```

It was labelled "The Aero Broadcast Seed Sower D.L.K.". and was inherited when Simon's grandad took over the farm in 1947. It's a wonderfully quirky machine, designed to scatter grass and clover seed as the sower walks through a field. And you do that by filling a canvas bag with seed and then moving a bow at the front of the machine back and forth. This drops seed onto a spinning disc, which then scatters it evenly in a wide arc up to 6m (20ft) across. It's called a "fiddle drill" because the back and forth motion of the bow is much like a musical fiddle. Simon actually thought his dad was playing some kind of musical instrument when he saw him using the fiddle drill for the first time.

I used my trusty circle-cutting machine (see page 112) for the first stage of remaking the spinning disc for the fiddle drill. The second stage involved one of the most complex soldering jobs I'd ever attempted (I'd never actually done any soldering in my life before I joined The Repair Shop, believe it or not) but I learned a lot from Brenton West (our silversmith) and Pete Woods (our musical instrument expert) on the show, who were both so willing to share their knowledge with me.

The Aero brand was produced in Kilmarnock, Scotland, but the fiddle drill came from the US, where it had been around since at least the 1860s. The most famous was made by the Cyclone Seeder Company, set up in 1868 by Samuel S. Speicher in Indiana. By 1890, it was producing up to 40,000 a year and had established an international trade office in Queen Victoria Street, Central London.

I'D NEVER ACTUALLY DONE ANY SOLDERING IN MY LIFE BEFORE I JOINED *THE REPAIR SHOP*, BELIEVE IT OR NOT.

FIDDLE DRILL
This piece of machinery is used to scatter seeds across a field, but it shares its name with the musical instrument because it is operated by using a bow in a similar fashion to fiddlers.

FAIRGROUND AND NOVELTIES

MACHINE NUMBER	MACHINE NAME	PAGE NUMBER
074	CANDYFLOSS MACHINE	170
075	POPCORN MAKER	172
076	BATHING MACHINE	174
077	TEST YOUR STRENGTH MACHINE	176
078	FORTUNE-TELLING MACHINE	178
079	PINBALL	180
080	CAROUSEL	182
081	MUTOSCOPE	184
082	MUSIC BOX	186
083	PEDAL CAR	189
084	BOOK-VENDING MACHINE	190
085	PEDALO	192
086	SOAPBOX RACER	194
087	ROLLER COASTER	196
088	FERRIS WHEEL	198

CANDYFLOSS MACHINE

MACHINE NUMBER 074

This is another invention story that should be a Hollywood film. And possibly the most amazing thing is that candyfloss (cotton candy to Americans) was actually invented by a dentist: William J. Morrison, from Nashville, Tennessee, who began working on a machine to spin sugar in the 1890s. And he wasn't just any old dentist, either, he was President of the Tennessee State Dental Association, no less.

He eventually teamed up with confectioner John C. Wharton, also based in Nashville. On the patent, which they filed together in 1897, they just called it a "candy machine". It worked by melting sugar crystals in a large metal bowl, perforated with tiny holes. A rotating cylinder (operated by a foot treadle) then span rapidly, forcing the liquefied sugar through the holes, and as it travelled through the holes, it cooled, solidified and formed long strands.

Spun sugar has been a thing for hundreds, if not thousands, of years, but seeing as sugar was very expensive in the Middle Ages, it was something you'd only really see at a royal dining table. There is a famous tale of French King Henry III visiting Venice in 1574, and being presented with plates and silverware made from spun sugar. Designs got increasingly elaborate in the 19th century, with the renowned French chef Marie-Antoine Carême (who created Napoleon Bonaparte's wedding cake) even creating sugar sculptures of windmills, fountains and temples.

Morrison and Wharton had created an eye-catching product, but what they needed was a big event so that they could wow as many people as possible. And as luck, or design, would have it, the same year that they applied for the patent, the Tennessee Centennial Exposition (to celebrate the 100th year of the state's founding) came along, attracting around two million people. But it was the St Louis World's Fair in 1904 that took the duo's sweet treat stratospheric. That event was attended by almost 20 million people and

POSSIBLY THE MOST AMAZING THING IS THAT CANDYFLOSS (COTTON CANDY TO AMERICANS) WAS INVENTED BY A DENTIST: WILLIAM J. MORRISON.

featured exhibits from over 60 countries. It brought other products like peanut butter, hot dogs, hamburgers and waffle-style ice-cream cones to a huge international audience.

Morrison and Wharton called their spun sugar "fairy floss", exhibited it in the fair's "Palace of Electricity", and presented it to customers in little wooden boxes. Over the course of the seven-month event, they sold 68,655 servings at a $0.25 a pop. That works out as over $17,000 – equivalent to over half a million dollars in today's money.

The only thing that didn't stick was the name "fairy floss", except in Australia. It became known as cotton candy only in the 1920s, thanks to another dentist turned confectioner by the name of Josef Lascaux, who developed a different machine in 1921. Amazingly, he sold his cotton candy from his dentist's surgery. The product became known by some pretty amazing names around the world – in the US, it's cotton candy; in the UK (and much of the Commonwealth), it's candy floss; in France, it's called *barbe à papa* (daddy's beard); in the Netherlands it's *suikerspin* (sugar spider); and in the Afrikaans language, it's *spookasem* (ghost breath).

While most cotton-candy machines are now fully automated, and can spin at around 3,500 revolutions per minute, you still find some around that operate via a foot treadle. And I'm a sucker for anything that you operate with a treadle, using your own steam.

CANDYFLOSS MACHINE
Although most candyfloss machines are now fully automated, they were originally operated by a foot treadle.

POPCORN MAKER
Charles Cretors purposefully designed his popcorn maker to combine engineering with street theatre to draw a crowd.

POPCORN MAKER

MACHINE NUMBER 075

There's an urban legend floating around that popcorn was served at the first Thanksgiving feast back in 1621, but it wouldn't have been possible, given that the corn that they grew at the time in the area was too fragile to make popcorn. It's a good story, though, and not actually that far-fetched.

We know that Indigenous Iroquois people, who have inhabited the areas of northeast North America for over 4,000 years, were popping corn kernels using heated sand in pots from at least the 1610s, and used it to make a type of soup, but we don't know how long they'd been making it for. Popcorn as a snack became popular among the European settlers over the next 200 years. Some colonial families even ate it for breakfast, with cream and sugar, but it was very much a make-it-at-home-in-a-basket-over-the-fireplace kind of operation.

Most popcorn is now made at home by heating up a slightly tragic bag in the microwave, but it's a far cry from the excitement you would have felt seeing and smelling it for the first time at a fair in the 19th century. The person who made that happen was Charles Cretors, a painter and decorator turned confectioner based in Decatur, Illinois.

Keen to generate a bit of a buzz for his shop, he purchased a commercial peanut-roaster so customers could buy freshly roasted nuts. Unhappy with the results, he sold up and relocated with his family to Chicago to start work redesigning the machine for roasting nuts and for popping corn. He changed the power source to a very small steam engine, and worked on making the whole process more dramatic, to attract an audience, even adding a figurine of a mechanical clown (which he called Tosty Rosty) to his new machine, and devising a transparent case so everyone could see the steam engine. It must have been pure street theatre and a great combination of showmanship, advertising and engineering. To be fair, if I saw someone selling any type of food powered by a steam engine, I think I'd buy it. And product-wise, the results were much better – the kernels were heated much more evenly, so they tasted better and he'd end up with fewer wasted unpopped kernels. Cretors applied for a peddler's licence in December 1885 so he could stick his new machine outside his new shop. This marked the beginning of C Cretors & Co.

In a similar fashion to Morrison and Wharton's electric candy machine, Charles Cretors sought a stage, and he got one at the World's Colombian Exposition (so-called to celebrate the 400th anniversary of Christopher Columbus's voyage to the Americas) in 1893, conveniently taking place in his now-home city of Chicago. By that time, Cretors had started popping the corn in a special seasoning which coated every kernel the same way. His little steam engine powered the fan blades at the bottom of the device to continually stir the popcorn while it cooked. His popcorn was a hit at the Exposition. Seven years later, Cretors had designed a horse-drawn popcorn wagon, and the company had moved into large offices in central Chicago where they remained for 76 years. They're in Chicago and owned by the Cretors family to this day.

BATHING MACHINE

The bathing machine was basically a large, covered, wheeled cart — about 1.8m (6ft) high and 2.4m (8ft) wide — to protect people's modesty before they entered the water. Seeing as it was customary to swim naked, the bathing machine was usually wheeled up to or into the water by a horse and cart. Then the occupant would walk down the steps and into the water.

They'd often have a little flag, which they'd raise to alert the horse and cart that they were ready to return to shore. As for when they first appeared, Scarborough Public Library has an engraving by draughtsman John Setterington believed to date to 1735–1736 which shows a bathing machine (or at least a horse-drawn box) in use at Scarborough beach. It has four cart-like wheels, all the same size, a pitched roof and a window. The naked occupant is walking out at the water's edge as the floor of the box is quite low. The folks down in Devon also claim the first bathing machine appeared in their county in 1735, although I haven't seen an illustration of their one.

In 1750, Margate resident Benjamin Beale, a glove-maker and a Quaker, developed a more sophisticated bathing machine. This was one with much larger rear wheels than front ones, a much higher-set hut, wooden steps leading down several feet into the sea, and a canvas awning that was later referred to as a "modesty hood". The machine was designed to be driven by horse and cart actually into the sea and then you'd walk out and take your dip.

As for bathing suits, men were allowed to swim naked until 1860 when it was banned. It was a different story for ladies, though who, from the middle of the 18th century, were wearing ankle-length woollen dresses with massive sleeves and black stockings so that their shape wasn't visible. Not exactly your optimum swimming attire.

The most famous bathing-machine users were Queen Victoria and Prince Albert. The royal couple used their machines to swim in the sea near Osborne House – their spectacular summer home on the Isle of Wight. And Victoria's bathing machine was a pretty spectacular contraption too, featuring a verandah at the front, curtains and a plumbed-in

AFTER QUEEN VICTORIA DIED, HER BATHING MACHINE WAS USED AS A CHICKEN COOP, BUT IT WAS RESTORED IN THE 1950S.

toilet! After Victoria died, her bathing machine was used as a chicken coop, but it was restored in the 1950s and now stands proudly on the edge of Osborne beach.

Modesty became much less of an issue around the turn of the 20th century and from 1901, men and women didn't have to occupy different areas of the beach. Bathing machines became obsolete, and many lost their wheels and were turned into beach huts, which became a more permanent fixture of the British seaside, but bathing machines haven't all died a death. A beautiful new one was built in Margate in 2014, an idea devised by local skincare-store owner Dom Bridges. He kick-started a crowfunding campaign and got together local craftspeople to work on the construction of the bathing machine. And now, not only is it a beautiful contraption, with its giant oak wheels and slatted black pine body, it's also a fully functioning sauna. Known as the Haeckels Community Sauna, it's also completely free to use.

BATHING MACHINE
To protect their modesty, people used to be taken either right to the water's edge, or even into the water, in their bathing machine, before descending the steps for a dip.

TEST YOUR STRENGTH MACHINE

MACHINE 077 NUMBER

This fairground classic was also known as the Strongman or High Striker machine, and involves whacking a pad or a cushion with a long-handled heavy mallet as hard as you can.

This sends a weight upwards inside a vertical wooden tower containing a scale reaching from 0 to 100. The heavier the hit, the closer you'll get it to 100, and if you do that, the weight rings the bell at the top. Some of these towers are only a metre or so high, but others can top 6m (20ft).

One of the companies that developed a high striker machine was the Gatter Novelty Company of Philadelphia. It was formed in 1923 by two brothers, Richard and Rudolph Gatter, and they made coin-operated fairground machines for amusement parks in Coney Island, New York, and Atlantic City, New Jersey. The company struggled after the devastating Wall Street Crash of 1929, and the brothers sold the business to the Exhibit Supply Company of Chicago, one of the most famous penny arcade manufacturers in the US. Gatter also made a striking clock strength-tester machine, where the player has to press two bars together, release them and see how many times the clock strikes.

In the UK, Carters Steam Fair, which was run by the hugely talented signwriter, gilder and restorer Joby Carter until late 2022, had three "strikers". You might have seen them (and other Carters rides) appear in films over the years, including *Paddington 2* and *Rocketman*. The oldest of the three strikers, called the Mighty Striker, was actually the first item of fairground equipment bought by Joby's family, in 1975. It dates back to 1948 and was built for legendary fairground showman Chicken Joe. (Joe earned his name from his stall, which involved buying a raffle ticket and seeing if its numbers came up on a big spinning wheel. If they did, you'd win a bag of groceries with a chicken on top). Joby built the other two strikers himself, the Super Mother Striker and Son of Striker, for smaller folks wishing to test their strength. Until I got to know Joby, I had no idea that on Test Your Strength machines, someone has to actually catch the weight that goes up the column as it plummets back downwards. It's quite a high stakes operation, because if you don't catch it, it'll smash the bottom of the machine. It's just one of those things that takes place behind the scenes at a fairground, but now you know to look out for it.

One of my favourite ever *Repair Shop* fixes was a unique Try Your Strength machine. It weighed about a ton and was one of the trickiest things we've ever shifted into *The Repair Shop*. It must have looked like a comedy sketch with Jay Blades and me doing a lot of "to me, to you" while we attempted to manoeuvre it over the gravel surrounding the barn. It was brought in by a lady called Marie; she'd inherited it from her late husband, Rob, who'd fallen in love with it after seeing it in a café when he was out cycling with his dad back in the 1950s. He went back to the café in the 1970s, and saw that all their penny amusements were up for sale, so he bought it. He was a collector and a fellow restorer. He sounded like a man after my own heart, and I wanted to do his memory justice.

The machine had been outside for a long time, exposed to the elements, so it had seized up. I ended up cleaning every one of its components, but it's when you really invest the time and effort into a treasured object like that, that you're able to discover its secrets. And one of them was the original pea-green colour it was painted in, hidden beneath the semicircular Try Your Strength sign. We had the cast-iron frame sandblasted while I was rebuilding some of the fixings that were beyond repair, and then it was ready to paint. You often have to use a certain amount of artistic licence with colour because while I found that little pea-green section that I could colour-match from, I know that other sections of the machine were likely to be different shades of green. You just have to go with what you think looks good and respects the original design.

I remember having a huge pile of tiny screws that all needed to go back into this jigsaw puzzle of a mechanism, but there's nothing like it, when everything's gleaming and working perfectly. The actual dial for the machine was made by A. Rowland & Son, based at Smedley Road works in Manchester. The cast-iron frame was made by the amazingly named Improved Strength Testing Machine Company, and they were obviously a quality outfit because Queen Victoria's crest was visible on the frame, indicating that they'd been awarded a royal warrant. I added gold leaf to the raised lettering to give it a finishing flourish. I was actually really nervous (as well as excited) when we unveiled it, but Marie loved it. And she was happy for *The Repair Shop* gang – who'd all been itching to give it a go – to test their muscles on a unique piece of fairground fun.

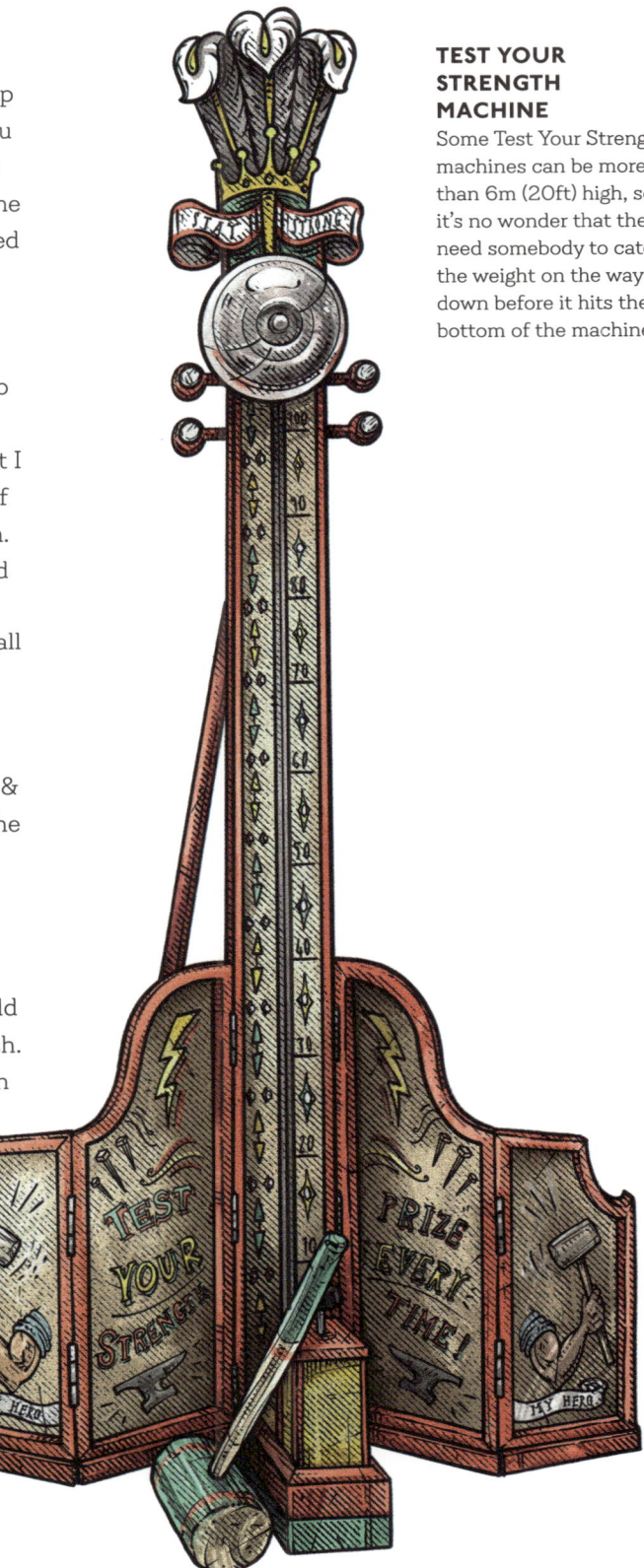

TEST YOUR STRENGTH MACHINE
Some Test Your Strength machines can be more than 6m (20ft) high, so it's no wonder that they need somebody to catch the weight on the way down before it hits the bottom of the machine.

FORTUNE-TELLING MACHINE

MACHINE NUMBER 078

```
This is one of those iconic machines that takes me back
to the seafront amusement arcades in Southend, Essex
when I was a kid. I'd go in with my mates after grabbing
a cheeky hot dog or a doughnut — the ones that came out
of those conveyor-belt machines. There's nothing like the
smell of freshly made doughnuts. All the shops on the
seafront had those big bags of candy floss hanging up.
```

I remember putting a £1 coin into a change machine and feeling like I'd won something when the steady stream of coppers came out, then you'd scoop them up into one of those little ice-cream pots and head over to the 2p machines. We'd be there for hours until all the coins were gone. It didn't matter that you never won anything — to be fair, I don't think anyone's ever left with more than they started with.

The Southend Pier is amazing. It's the longest pier in the world, and also one of the oldest still functioning. The iron pier that's there now opened in 1889, and it replaced an even older timber one. It has survived at least four major fires and it's still going strong. It has even got a pier railway — the first of its kind in the UK — that runs from the shore to the pier head. The gaps between the boards on the pier walkway are slightly frightening, and they seem to get wider every time I see them, but it's an incredible feeling walking more than 2km (1.2 miles) out into the Thames Estuary. You really do feel like you're out at sea.

Next to the pier is the Adventure Island theme park, but everyone who grew up in Southend remembers it as Peter Pan's Playground. There was also an area known as Never Never Land opposite the seafront, with fibreglass models of obscure cartoon characters, goblins and fairies around magical castles. It was like a sort of low-budget *Alice in Wonderland* set.

Around the pier were so many neon-signed amusement arcades, I remember wondering how they all stayed in business. And there was one machine with a name that anyone who grew up in the 1980s will know: Zoltar.

Zoltar was made famous around the world after the 1988 film *Big*, when 12-year-old kid Josh Bashkin (played as an adult "big" child by Tom Hanks) makes a wish to become "big" on a fortune-telling machine called Zoltar Speaks.

The first of this kind of machine dates back to 1867, believed to be invented by an Englishman known as J. Parkes. You put a coin in the slot and several discs started to turn, printing your fortune out on to a little ticket.

In the US, one of the first fortune-telling machines was created by Mills Novelty Company of Chicago, one of the biggest producers of coin-operated slot machines in the country. It was called The Wizard Fortune-Teller, and was made in 1900. The user put a penny in the slot, and moved a metal dial towards one of the six boxes, which contained a question they wanted answering such as: *What is my principal quality?*; *What is my*

greatest defect?; How many times shall I marry?; Shall I soon fall in love?; My future occupation (there were two boxes for this question, one for men and one for women).

Mills also produced a talking Gypsy fortune-telling machine in the first decade of the 20th century. One, thought to be one of only three left in the world, became the property of the state of Montana in 1998. They went about restoring it in 2004, and it was the subject of press coverage after it attracted the attention of magician David Copperfield, who'd been collecting early penny-arcade machines for years. Ultimately, the machine was restored to its former glory, and, in 2008, was installed in the Gypsy Arcade, in the town of Virginia City, Montana. To play, it costs a nickel, which travels down a track and moves a lever, which closes a switch, activating a motor and a solenoid. The motor moves a belt, which starts moving the head, mouth and eyes, and a concealed record-player speaks your fortune. Frankly, it's terrifying.

FORTUNE-TELLING MACHINE
Well recognised by anyone who has seen the film *Big*, Fortune-Telling Machines were a common sight in amusement arcades, although frankly terrifying to actually consult!

PINBALL

MACHINE NUMBER 079

Pinball's origins can be traced all the way back to miniaturised indoor versions of lawn bowling, which was very popular in Elizabethan England. By the 17th century, this had become an indoor table-game that involved aiming a small ball with a cue towards target areas protected by a circular arrangement of wooden pegs.

This game eventually became standardised as the game bagatelle, thought to be named in 1777 after the Château de Bagatelle where French King Louis XIV attended a party where the game was a feature. Sometime between the 1750s and 1770s, a similar game was invented which became known as *billard japonais* (Japanese billiards). It used a spring mechanism to fire balls from a channel to the top of an inclined board, and down towards the various target areas. Versions of this game were beautifully designed, resting on four carved and shaped wooden legs, and featuring little circular stands that swung out of each side of the counter to hold candles.

In 1871, the British inventor Montague Redgrave, who had emigrated to Cincinnati, Ohio two years before, patented his own variant of Japanese billiards. It was a smaller, portable game that you could put on a countertop, and featured 14 target areas, including two larger, upside-down keyhole-like openings with a bell in the centre. Like Japanese billiards, you launched a ball onto the board via a spring-loaded plunger, but it used clay marbles instead of ivory balls. Redgrave's game was a big success, and is widely regarded as the precursor to pinball.

By the 1930s, the coin-operated bagatelle machine had been created. The first hugely successful commercial version, *Baffle Ball,* was released in 1931 by Chicago company D. Gottlieb & Co, founded by David Gottlieb in 1927. The game dispensed five balls for a penny and went on to shift over 50,000 units at $17.50 each, helping to establish D. Gottlieb & Co as one of the biggest manufacturers of coin-operated games in the US. In 1933, engineer and inventor Harry Williams designed the first electro-mechanical bagatelle machine, which was called *Contact*. Williams installed electromagnets, called solenoids, which first propelled the ball onto the board (hence the name of the game) and then manoeuvred the ball from low-scoring holes into higher-scoring holes at the bottom of the board, if you shot the ball over a switch at the top of the board. Pinball wasn't coined as the name until 1936, the same year that the first "bumper" was created, by inventor W. Van Stoeser in the bowling-themed game *Bolo*.

In 1947, Gottlieb released the pinball game *Humpty Dumpty* – the first pinball machine to feature flippers, invented by designer Harry Mabs, to help keep the balls in play. But the 1940s wasn't a happy time for pinball – its perceived links with gambling and crime, and the concern that kids were both skipping school and spending their money on pinball rather than lunch, meant that it was banned in several major American cities including New York, Chicago and Los Angeles. It was only in the 1970s that the bans were overturned, and while that coincided with microchips replacing electromechanical parts, it was also a time when pinball machines faced huge competition to their very existence: arcade video games. But since the 1990s, pinball has made a surprising comeback. According to the International Flipper Pinball Association, pinball competitions have increased five-fold in the last decade. And perhaps more pleasingly, craftspeople are repairing old machines left in dusty storerooms and broken-down warehouses. And that's no easy task, I can tell you.

It's mind-blowing how many wires there are when you lift up the cover of a pinball machine. It looks like the inside of one of those telephone exchange boxes you see on street corners in the UK. There are hundreds of wires, lights, levers and electromagnets. And then you've got the stunning artwork, which is a whole other side to pinball machines. The attention to detail on every aspect of a pinball machine is incredible. And I've been lucky enough to get a first-hand education from a chap whose pinball knowledge is unparalleled: Geoff Harvey, better known as "Pinball Geoff".

Geoff is a *Repair Shop* expert, and one of the jolliest, most positive people I've ever met. He has been a massive pinball fan since he first played a machine on a trip to Jersey. A few years later, he took advantage of his parents being out, and shifted ten pinball machines into their house. His mum came back to find that she couldn't actually open the front door, because there was a pinball machine bottleneck in the hallway. After that incident, his parents made it clear that he could collect pinball machines, but they'd all have to go in his room. So, of course, Geoff got rid of anything in his room that he didn't need – including his bed – in order to store his pinball machines. He has become really involved in the Glastonbury-on-Sea area, near the Healing Fields, at Glastonbury Festival that has been going since 2019. This is basically a full-sized recreation of a British seaside pier (albeit with a quirky Glastonbury twist), with wooden boards, Victorian-style railings, deckchairs, lampposts, lifebuoys, a postbox, a Punch and Judy show… and a fantastic pinball alley, run by Geoff. I've never owned a pinball machine but I would love to one day, and the more time I spend with Geoff, the more likely that becomes. The closest I got was playing *Pinball Dreams* on the Amiga in the mid 1990s with my brothers. The question is, though: which pinball machine would you go for? A *Back To The Future* one? *Jurassic Park* maybe? You can see why I often end up down an eBay rabbit hole.

PINBALL
From bagatelle, to Japanese billiards, to today's pinball machines, there has long been an attraction to firing a ball and gently persuading it to drop down a high-scoring hole.

CAROUSEL

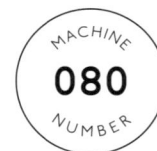

MACHINE NUMBER 080

The steam-powered mechanical roundabout (also known as a merry-go-round) was invented by Englishman Thomas Bradshaw, and first shown to the public on New Year's Day in 1861 at Bolton Pot Market, Lancashire.

The engine was constructed at Pollit's Boiler Yard in Bolton, and built by Messrs Rogerson and Brimelow of Deansgate, while Bradshaw himself made the beautiful wooden horses.

It was described in the local newspaper, the *Halifax Courier*, as a "roundabout of huge proportions, driven by a steam engine which whirled around with such impetuosity, that the wonder is the daring riders are not shot off like cannon-ball, and driven half into the middle of next month." This was at a time when members of the medical profession had serious concerns about what speeds of more than 50km/h (30mph) would do to the human body and brain.

It was the renowned English fairground-engineer Frederick Savage who made some innovations to the roundabout design that we know and love today. In 1885, he built his *Platform Gallopers* which gave the horses their up and down motion (to mimic galloping) thanks to a series of "eccentrics" (circular discs fixed to a rotating axle, but the centre of the disc is at an angle to the axle, hence the name "eccentric", which literally means "out of centre"). Savage also came up with a lateral sliding mechanism that enabled the horses to slide outwards by up to 15° when the roundabout had picked up enough speed (and therefore centrifugal force). Also in 1885, Messrs Reynolds and King came up with the overhead crank system, suspending the horses from above, which Robert Tidman & Sons of Norwich adapted and improved the following year. The oldest ride owned by Carters Steam Fair is the Jubilee Steam Gallopers, built around 1895 by Robert Tidman & Sons. It features a 46-key organ in the centre, made around 1900 by renowned Parisian manufacturers Gavioli, and no carousel ride is complete without the atmospheric, iconic piped-sound of a fairground organ.

In the UK, the Gallopers ride turns clockwise and features very similar-looking horses, but in the US and continental Europe, a carousel (which became the popular name there for a merry-go-round at the end of the 19th century) turns anticlockwise and the horses can look quite different to each other. The oldest operating carousel in the US is the Flying Horses Carousel on Martha's Vineyard, an island off the coast of Massachusetts. It's thought to date from around 1876 and was designed by Charles W. F. Dare Company. It was originally powered by a steam engine, but was converted to run off electricity in 1900. The carousel still has a brass-ring device that dispenses a single brass ring at the end of a wooden, or metal, arm as the carousel turns. You can just about grab it if you lean from the outside ring of horses, and if you do, you win yourself a free ride on the carousel.

The carousel is absolutely the heart of the fair. It's always the most highly decorated ride, and the amount of work that goes into it is extraordinary. I say that as a guy who repaired just one carousel horse back in Series 4 of *The Repair Shop*. So many different skills are on show when you look closely at a carousel. You need an expert in cut glass, brilliant cutting, signwriting, gilding and carpentry – and that's before you even consider the epic engineering behind the pipe organ and the steam engine. And all of those skills come together to create something extraordinary which puts a smile on everyone's face. It's such an immersive sensory experience, smelling the steam engine, hearing the fantastical notes of the pipe organ, seeing the up-and-down motion as the lights, poles and horses whizz by. It always makes me think of the people who have been on it before and the joy that it's brought them. It's a cultural treasure.

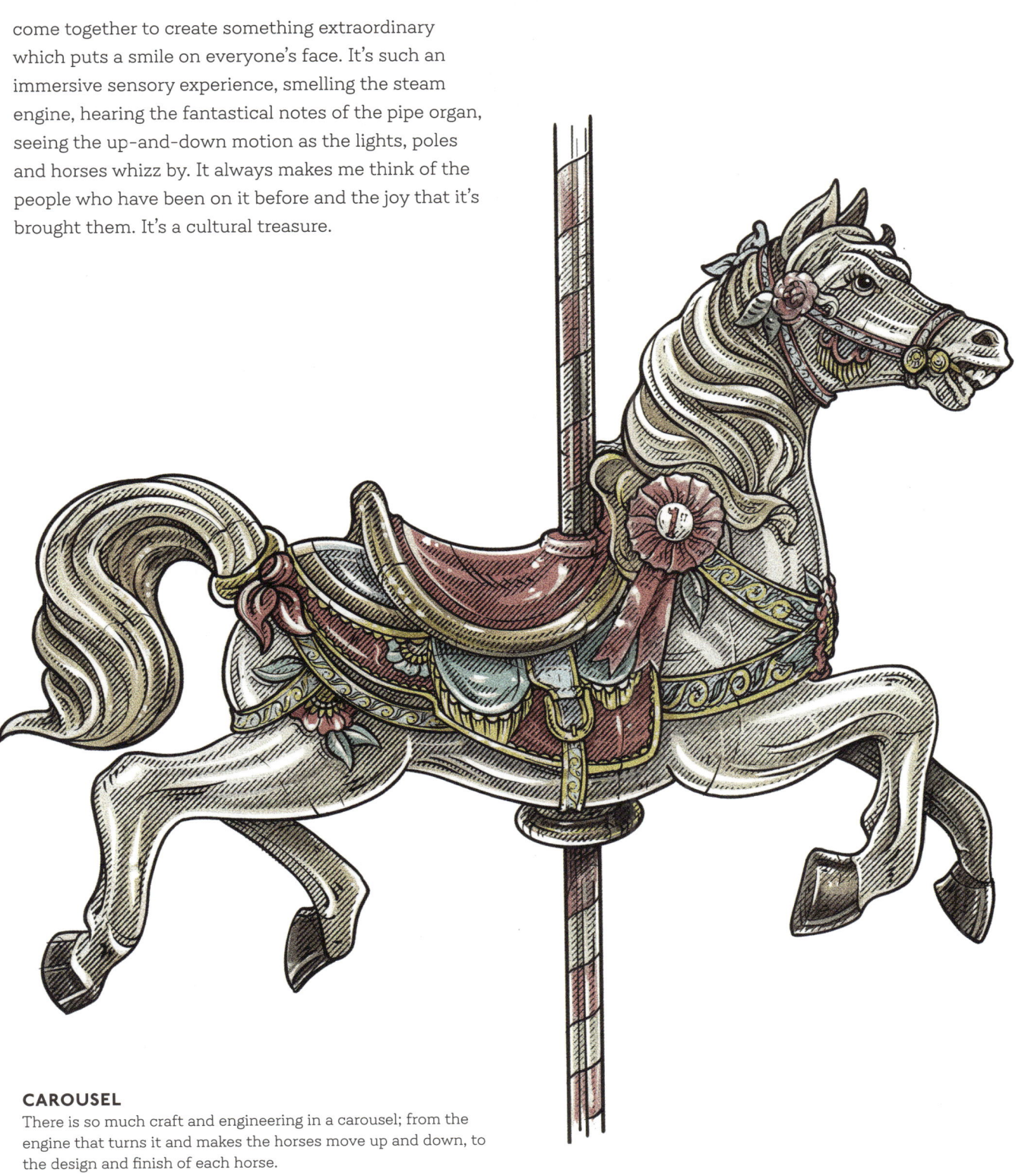

CAROUSEL
There is so much craft and engineering in a carousel; from the engine that turns it and makes the horses move up and down, to the design and finish of each horse.

MUTOSCOPE

Another classic feature of an amusement park and pier was the Mutoscope. The Mutoscope was an early motion-picture device containing hundreds of sequential photographs, arranged much like a really big Rolodex, and when the user operated the crank, they turned so quickly that they seemed to move.

The American inventor Herman Casler filed a patent for the device in 1896, but the invention was a collaboration between Casler and the British inventor William K. L. Dickson, who had emigrated to Virginia and was hired at Thomas Edison's laboratory in 1883. Dickson had been largely responsible for developing Edison's Kinetoscope – another early motion-picture device that had been unveiled to the public in 1894. Unlike the Mutoscope, there was a continuous loop of film which ran inside the Kinetoscope on rollers powered by an electric motor, but the effect was very similar, creating the illusion of moving pictures.

The Mutoscope was simpler and sturdier than the Kinetoscope, and was manufactured from 1895 by the American Mutoscope Company, the firm Dickson formed with partners Casler, Henry Norton Marvin and Elias Koopman. The device was licensed to William Rabkin in 1920, who had founded the International Mutoscope Corporation. From the beginning, Mutoscope reels were notoriously "racy" for the time. They became known as "What the Butler Saw" machines in the UK, after the title of one of the famous reels produced for the device. It was inspired by a high-profile divorce case from 1886, involving Lord Colin Campbell and his wife Gertrude Elizabeth

MUTOSCOPE
These early motion-picture devices, operated by a crank handle, were known in the UK for their slightly racy films.

Blood. Members of the jury were invited to their house in London's Belgravia to scrutinise testimony from the butler over whether or not he would have been able to witness adultery by looking through the keyhole of their dining room.

The iconic devices, often painted red, change hands on eBay and auction sites for thousands of pounds. You can find them in seaside museums across the country, but the one I really got to know is housed in Brighton's fantastic Toy Museum, nestled under the railway arches beneath the station. And that was because a Mutoscope arrived on my workbench on *The Repair Shop* a few years ago. Well, it was half a Mutsocope really, because it was missing the stand. So I asked the guys at the Toy Museum if I could have a look at their one. And it was so useful to see one that had been made pretty much perfectly, because I'm not always so lucky with *Repair Shop* fixes. A lot of the time you're just working on guesswork, or your gut. But seeing it in the flesh made me realise how important the legs actually were to the overall design of a Mutoscope, with its X-shaped metal sides featuring a rose decoration in the centre. United with its newly made base, the Mutoscope I fixed stood proudly and instantly changed. You want to be able to see it from a distance and get excited by it, so giving it that kind of plinth, pedestal or stand elevates it not just physically but also psychologically. It lures you in. You're drawn to its beauty and that's part of the unique appeal of fairground and seaside attractions.

MUTOSCOPE
Illustration of a mutoscope dating back to c.1896, found in the Department of Photography at the Royal Institute of Technology.

MUSIC BOX

MACHINE NUMBER 082

The music box is thought to have been invented around 1770 in Switzerland, and was so small it could fit inside a pocket watch. Music boxes produce notes through a set of pins on a revolving metal cylinder that "pluck" the tuned metal prongs on a comb. You could buy different cylinders if you wanted a different tune, but changing them was a bit tricky.

To address that issue, by 1885, Gustav Adolf Brachhausen was producing "Symphonion" music boxes, near Leipzig, Germany, for the company Lochmannscher Musikwerke. The Symphonion was a mechanical music box that used metal discs rather than cylinders. The metal disc itself had punched slots that were "played" by ratchets. The whole contraption looked a lot like a very beautifully designed record player, housed in a square or rectangular wooden box, with a crank handle on the outside. In 1889, Brachhausen joined forces with engineer Ernst Paul Riessner and they founded Polyphon Musikwerke in Leipzig, producing their "Polyphon" music box as a competitor to the Symphonion. Some of the Polyphon designs were stunning, with cases made of walnut and featuring intricate floral marquetry. Brachhausen soon moved to the US with a small team of machinists and cabinetmakers to expand the business, and set up Regina Music Boxes in Jersey City, New Jersey, which became known as the Regina Company from 1902. It sold over 100,000 of its music boxes until 1921. Meanwhile, back in Germany, Polyphon Musikwerke had founded record label Polydor in 1913. A British subsidiary – Polydor Records Ltd – was set up in London in 1956, and signed some huge bands like The Jam and Dexys Midnight Runners. They're still one of the biggest record labels in the country, formed thanks to a new music box over 130 years ago.

MUSIC BOX
Original music boxes were so small they could fit inside a pocket watch, but that made it very tricky to change the tune. So, instead, new versions changed the design.

MUSIC BOX
Advertisement for the Stella music box by the Jacot Music Box Company, New York, 1901.

MUSIC BOX
Advertisement for Olympia music boxes by E. L. Cuendet, New York, 1900.

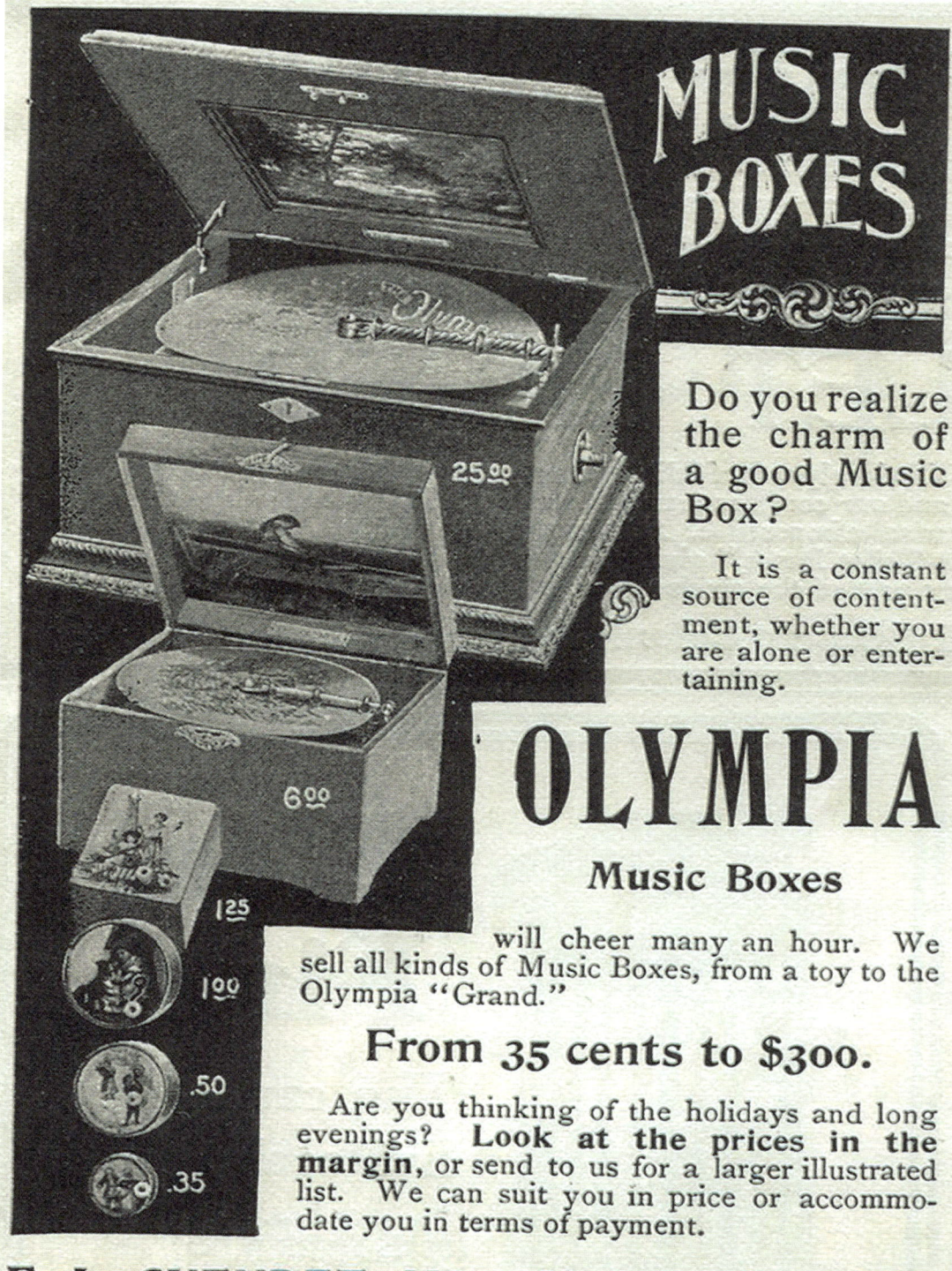

PEDAL CAR

The Austin J40 is the iconic pedal car, and I've loved it for as long as I can remember. It's a truly beautiful machine and has an incredible history. In 1946, the year that the Austin Motor Company resumed motor-car production after the Second World War, the boss came up with the idea of a pedal car for children.

Production began in 1949 and the design was based on Austin's A40 Devon and Dorset models. The little vehicles were made at the Austin factory at Bargoed in south Wales, and the UK government funded the project to employ disabled former miners. In fact, the entire workforce were ex-miners who'd contracted the lung disease pneumoconiosis in the pits, and had been registered as disabled. The employees worked with full-time medical supervision at the factory – it was the first operation of its kind in history.

The cars were produced using scrap metal left over from the production of Austin cars. The build and design quality was extremely high – the cars featured working headlights, a horn, detachable wheels with pneumatic tyres, a bonnet that could open, chrome bumpers, leather upholstery, hub caps, a slatted grille, a boot handle, and a centre bonnet moulding complete with Austin's Flying A symbol. Demand for the J40 ("J" stood for Junior) grew – a young Prince Charles was even given one – and so did the workforce, reaching around 500 people. Austin kept producing the much-loved J40 until 1971. They made 32,098 of them at the factory, and many were exported to the US and Canada.

The pedals on a J40 are offset (i.e. one's forward and one's back) and connected to a rod that pivots. As you push each pedal forward, it rolls a crank over that's connected to the rear axle. So you're basically using a *push, push, push* motion that's very similar to the treadle of a sewing machine, only you've effectively got a double treadle.

In recent years, there's been huge interest in J40s. And that has everything to do with the sweetest car-race around: the Settrington Cup at Goodwood Revival. It's for kids, just features Austin J40s, and everything about it is wonderful. It does have a

I MANAGED TO GET HOLD OF A COMPLETELY KNACKERED, RUSTED, ROTTEN ONE, WHICH IS CURRENTLY HANGING ON THE WALL IN MY WORKSHOP.

competitive edge, though, and all cars have to go be scrutinised to make sure there haven't been any sneaky modifications undertaken by any uber-competitive parents. The cup is awarded over two races, and grid positions are allocated by pulling numbers out of a hat, which are then reversed for the second race. The races themselves begin, Le Mans-style, with the drivers running across the track to their cars and then, they're off. It's all over in a few minutes, but it's an incredible spectacle with a lovely sense of camaraderie among the competitors and parent-mechanics. A lot of the kids are the children of racing drivers.

I'd been trying to get hold of a J40 for years, but since their revival at Revival, their prices have skyrocketed. I managed to get hold of a completely knackered, rusted, rotten one, which is currently hanging on the wall in my workshop to remind me to get on to it as soon as I can. And I will do, especially because I've got to know the owners of the J40 Motor Company, which was established in 2020 to supply parts and accessories for J40s. In 2022, the owners bought the Austin trademark, which means they could produce new pedal cars under the Austin name. And so Austin Pedal Cars was born. They revealed a new J40 prototype continuation car at Goodwood Revival in 2022, and the completed production model was finished in October 2023. Exciting times.

BOOK-VENDING MACHINE

MACHINE NUMBER 084

Amazingly, there was an early vending machine in 1st century Ancient Greece (although they may have even existed earlier than that). Hero of Alexandria (what a name that is), fabled mathematician, engineer and inventor of the first steam-powered engine, described a machine where a user inserts a coin to receive a small quantity of holy water in a temple. When the coin was inserted, it tipped a balance, raising a plug in an urn and letting water escape.

In the 18th century, you could buy snuff (and later cigars) from a small rectangular brass box. You insert a half-penny and pressed a button, which allowed you to access the entire box of tobacco, so it was up to you to take the quantity you'd paid for. It's for this reason they become known as "honesty" or "honor" boxes. There is one obvious problem with such an object, though, as the choice of location was a tavern.

The book-vending machine was invented by tinsmith turned political activist Robert Carlile in 1822 in England. As with the tobacco-vending machine, it was a good plan... up to a point. Carlile reckoned that if you sold subversive pamphlets and banned-publications through a machine, rather than from a person, the seller would escape punishment. He was wrong. Although, it was his friend who ended

up in jail, which he probably didn't thank his employer much for.

The most famous book-vending machine was the Penguincubator, created by Allen Lane, who founded Penguin Books, in 1935, with his brothers John and Richard. The story goes that, in 1934, Allen was returning from Devon on a trip to visit author Agatha Christie, and found himself at Exeter St Davids Station without anything to read for the journey home. There was a bookstall at the station but it sold magazines and Victorian reprints. That was the moment when the good-quality, reasonably priced paperback was created – in his mind, at least. In 1937, a Penguin paperback-vending machine, dubbed the "Penguincubator", appeared outside the bookshop Henderson's, at 66 Charing Cross Road, in London's West End. Penguin RandomHouse installed another book-vending machine at Exeter St Davids Station in 2023, in partnership with Great Western Railway and Exeter UNESCO City of Literature. The profits will help support the independently run Bookbag bookshop in Exeter and Exeter City of Literature.

PEDALO

Leonardo da Vinci drew a detailed sketch of a paddle-wheel boat around 1485. It was driven by a crank handle, which turned a large toothed-wheel that engaged a smaller wheel on the axle of a paddle wheel. But we can't say he invented it. Man-powered paddle-wheel boats had been used in China from at least the 8th century. It seems they were used as tugboats and to carry passengers across rivers.

Towards the middle of the 19th century, Victorian inventors started exploring the idea of a bike/boat hybrid, or a "water velocipede" as it was known. The first came about in 1868, and was in use on Lake Enghien just north of Paris, France. Similar devices were appearing in the US around the same time, and a number of water velocipedes were patented in 1869. There's an illustration of one in the incredible *Knight's Mechanical Dictionary* (published in 1882). It dates from around 1877, and is much more like a modern pedalo. It features a "driver" – looking a lot like Abraham Lincoln, complete with bushy beard, smart suit and top hat – who sits astride a large semicircular casing enclosing a paddle wheel, which he propels with pedals. He controls the steering through handlebars, which are attached to a cord on either side, and pulleys that connect to a T-shaped rudder. Two very elegantly dressed ladies carrying small parasols are seated in the front of the boat.

Georg Pinkert's Water Tricycle of 1891 became a big deal because its inventor, from Hamburg, Germany, used it to try and cross the English Channel from Cap Gris-Nez in the Pas-de-Calais, France to Folkestone, England. The machine was basically an oversized tricycle, featuring three big hollow wheels, with seven slim paddles fitted around the outside of the two rear wheels. An 1882 edition of the *Scientific American* described the journey: "A calm day was chosen, when Mr. Pinkert rolled his queer vehicle down the shore to the water's edge, and with the assistance of a man to push he worked out through

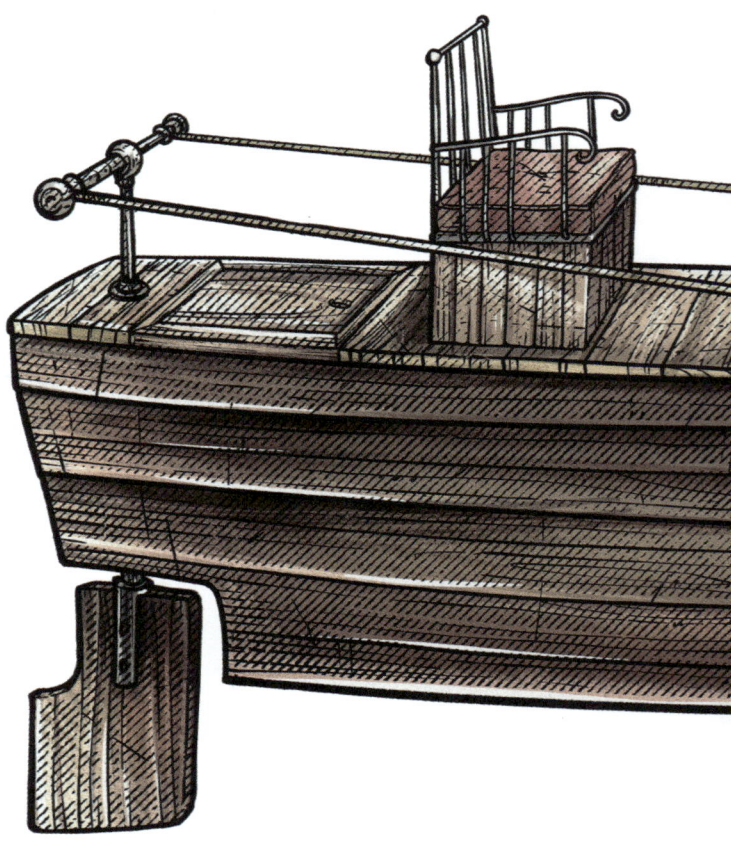

the breakers and headed for old England. It was pretty slow work but the inventor bravely continued his exertions. After many hours of labor and when half way across the tide turned and Mr. Pinkert became satisfied he would be carried away from land; so he hailed a passing vessel and was taken on board. He will probably make further experiments."

As for the classic pedalo, featuring two front seats fitted with pedals which drive a waterwheel under the vessel, that seems to have come about some time before the Second World War. There's an old Pathé News film of George Lansbury MP (grandfather of actress Angela Lansbury), the leader of the Labour Party from 1932 to 1935, opening Ravensbury Park in London in 1930, and taking two kids out "a-voyaging" on a pedalo. Everyone remembers going on a pedalo as a kid. For me, it was peddling a giant fibreglass swan pedalo, sometime in the 1990s, at the newly opened Lakeside Shopping Centre. I wonder what Leonardo da Vinci would have made of that.

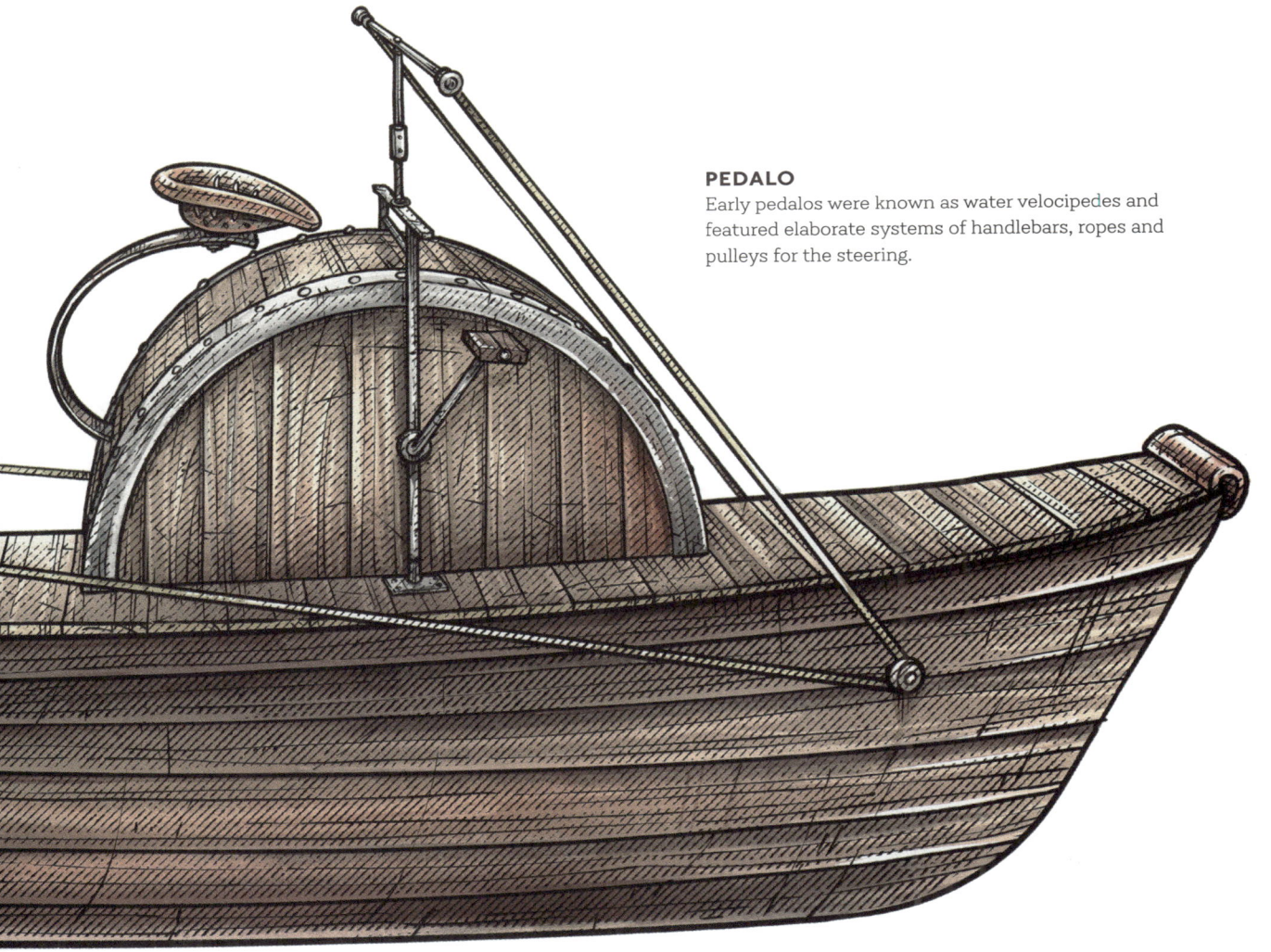

PEDALO
Early pedalos were known as water velocipedes and featured elaborate systems of handlebars, ropes and pulleys for the steering.

SOAPBOX RACER

MACHINE NUMBER 086

Soapbox racers have been an institution in the US ever since 10 June 1933, after a photographer had a brainwave. Myron Scott was the chief photographer of the *Dayton Daily News* and was asked to attend a race in Dayton, Ohio — famously the home of aviation pioneers Orville and Wilbur Wright — to cover a homemade pushcart race on a bumpy city street.

He thought it sounded like a good idea for the newspaper's Sunday Picture Page, but when he got there, he realised that this could be exactly the kind of good-news story that America desperately needed in the midst of the Great Depression. So Scott organised another race, for the following week, and told the boy racers to spread the word. The six "cars" in the first race were basically crates attached to buggy wheels, but for the second race, 19 entrants turned up with their weird and wonderful racers made of soap boxes, removal crates, barrels, fence panels, tin plate, and anything else hanging around in a garage or local junkyard. A large crowd turned up, and it seemed clearer and clearer that this was becoming a big deal. So Scott persuaded his bosses to sponsor the next race, which was to be a city-wide event set for 20 August 1933. Adverts were posted all over the place to encourage interest in a race featuring "anything on four wheels that will coast" on a sloped road to the east of Downtown Dayton. Over 450 contestants entered, although only 362 of the actual racers were allowed to compete, presumably because 90 of them looked extremely dodgy. Sixteen-year-old Randall Custer finished first with his "slashing yellow comet" three-wheeled racer, and won a motor scooter for his efforts. The third-place finisher was a surprise to almost everyone. "What's your name, son?" asked Scott before the racer removed a cap, revealing that he was standing in front of a 12-year-old girl called Alice Johnson. Amazing then that girls weren't officially able to race until 1971. That summer's day in 1933, 40,000 people turned up to watch. This had gone from local-interest segment to front-page news.

By the following year, the All-American Soap Box Derby Race had been set up, held on the same stretch of road as in 1933. The winner of the 1934 event, Bob Turner from Muncie, Indiana, was steering a car made from laminated wood that used to be a saloon bar. In 1935 the event moved to Akron, Ohio, and in 1936 a dedicated track was built, known as Derby Downs. The state Governor and Mayor of Akron opened the event, and the top prize was a $2,000 college scholarship. The Soap Box Derby is still held in Akron every year, and the track has three 3m/10ft-wide lanes and grandstands seating around 8,000 people.

Soapbox racing spread to the UK and became a feature of the Goodwood Festival of Speed between 2000 and 2004. I've been fortunate enough to spend some time with Charles Gordon-Lennox, the owner of Goodwood Estate, and the founder of the Festival of Speed and Revival. He showed me the private side of the house where his photo studio and that sort of stuff is. He had loads of black and white photos on the wall of some ridiculous-looking vehicles and he explained that they used to have a competition for kids at Goodwood Revival to design a car. The children came up with all sorts of weird and wonderful ideas –

double-decker, eight-wheels, cars with wings and propellers, that sort of thing – and then there was an area where engineers and mechanics actually built them out of scrap. It would be such a good idea to revive (again) – I'd love to get involved in that and then maybe the kids could race them around the track.

I made soapbox-style go-karts with my brother when we were kids. They were terrible and pretty lethal. But that didn't stop us from trying them out on our friend's garden, with a big downhill slope and loads of trees you had to chicane around. But I prefer that to the sponsored, professional-looking modern soapbox racers. I feel like soapbox racers need to be homemade, definitely feature pram-wheels, and look a bit ramshackle. That's part of the beauty of them.

SOAPBOX RACER
Soapbox racers became a big deal when a newspaper photographer spotted their potential to be the good-news story that Americans needed during the Great Depression.

I FEEL LIKE SOAPBOX RACERS NEED TO BE HOMEMADE, DEFINITELY FEATURE PRAM-WHEELS, AND LOOK A BIT RAMSHACKLE. THAT'S PART OF THE BEAUTY OF THEM.

ROLLER COASTER

MACHINE NUMBER 087

The roller coaster has its origins in 17th-century Russia, where sleds were ridden down specially constructed hills of ice, reinforced with wooden supports. These became known as "Russian Mountains". Apparently, Russian Empress Catherine the Great was a fan, and created a kind of summer version of this popular pastime by introducing wheeled carts that ran along grooved tracks down a hillside at her palace, in 1784.

In the late 1810s, sled rides appeared in amusement parks in Paris, thought to have something to do with the Russian soldiers who occupied the city soon after Napoleon's defeat at the Battle of Waterloo in 1815. The *Promenades Aériennes* (Aerial Strolls) was a ride that featured in the Parc Beaujon in 1817 on the Champs-Élysées, with wheeled cars fixed to a track with accompanying guide rails, but it was a temporary structure.

Then there was the famous Tivoli Gardens amusement park in Copenhagen, Denmark, which opened in 1843, featuring one roller coaster on a track that gave a "seven-second thrill" according to the park's website. In the US, one of the earliest and most famous roller coasters was Coney Island's Switchback Railway, which opened in 1884. It featured a tower, which riders would climb before boarding a wheeled car that reached a then-dizzying 10km/h (6mph) along a wooden track. But it was more of a scenic railway than a roller coaster, with passengers perpendicular to the track rather than facing it. The track took the car to another tower and then switched over to a return track, hence the name Switchback. Designed by LaMarcus Thomson, it was a massive hit. Amusingly, he designed it to offer people an excitable alternative to saloons and brothels which were appearing all over the place.

A big development came along in the 1910s with the invention of underfriction "safety" wheels, patented by US designer John Miller in 1919, which locked roller coaster cars onto their tracks. And with that sorted, all sorts of new thrills became possible, including higher top speeds and sudden turns.

And so began the golden age of roller-coaster rides. Amazingly, loop-the-loops had been the main feature of so-called "centrifugal railways", 75 years before the invention of underfriction wheels. It's not so much the risk of falling out that is alarming – gravity, momentum and centripetal force, which makes something follow a curved path, are the things that keep the car on the track – but rather the fact that these rides had completely circular loops (modern rollercoasters use much safer elliptical loops). That meant riders were experiencing G-forces not dissimilar from being involved in a car crash. Yikes.

ROLLER COASTER
It's thought that Catherine the Great was a pioneer of early roller coasters. Riding sleds on hills of ice was a winter pastime which she recreated on wheels in the summer time at her palace.

ALL SORTS OF NEW THRILLS BECAME POSSIBLE, INCLUDING HIGHER TOP SPEEDS AND SUDDEN TURNS. AND SO BEGAN THE GOLDEN AGE OF ROLLER-COASTER RIDES.

FERRIS WHEEL

MACHINE NUMBER 088

In 1889, the centrepiece at the Exposition Universelle, the World's Fair, held in Paris to commemorate the 100th anniversary of the Storming of the Bastille (and beginning of the French Revolution) was the Eiffel Tower.

At the time, it was the tallest building in the world and became a symbol of France. So you can imagine that by 1891, Daniel Burnham, the Director of Works in charge of the next World's Fair – the World's Colombian Exposition of 1893 – was getting pretty stressed about what the city of Chicago could come up with. He told his engineers to devise "something novel, original, daring and unique". Fortunately, one of the assembled minds – a 33-year-old bridge-and-tunnel engineer from Pittsburgh, who had founded a firm that tested metals used in civil engineering projects – came up with something. And his name was George Washington Gale Ferris.

Ferris's plan was to construct a gigantic 80m (264ft) steel wheel that rotated around a 13.7m (45ft) axle. It was to feature 36 suspended carriages that could accommodate up to 40 people each. He must have sounded like a lunatic when he pitched the idea, though it wasn't completely far-fetched. So-called "pleasure wheels" had existed in various cities across Europe and Asia from at least the 18th century, and lifted riders sitting in chairs or compartments, but these contraptions were basically made from wood and physically rotated by teams of strong men. Ferris's idea was on a completely unprecedented scale, involving 100,000 separate parts, and an axle larger than any that had ever been constructed before. But Ferris knew what he was talking about, and he won the contract to build it. Only, the deal was that he'd have to raise the finance himself – as if he didn't have enough on his plate. In all, the project cost approximately $385,000 (around $130 million today). And they started building it in winter, on completely frozen ground, which they had to blow up with dynamite before driving in 9.7m/32ft-long timber piles.

When the axle was made, by the Bethlehem Iron Company in Pittsburgh, Pennsylvania, it was the world's biggest hollow forging. It weighed 36 tonnes (89,320lb) and was attached to two cast-iron "spider webs" (the spoked framework of the wheel) weighing in at 21 tonnes (53,000lb). Then there were the two 43m (140ft) towers on which the axle turned. Supporting these two towers were eight 11m/35ft-high concrete pier foundations. Steam boilers drove the pistons of the two reversible 1000hp engines (one engine was there as a back-up). This fed steam through huge pipes to make the wheel turn. Air brakes, produced by Westinghouse Electric Corporation of Pittsburgh, controlled the movement of the wheel and held it stationary. The structure could cope in high winds, freezing temperatures and lightning strikes. Ferris had thought of everything and had produced a masterpiece of engineering. Even so, I can imagine the anxiety around the place when they performed a test run in June 1893. There were no cars attached at that point, but, amazingly, a number of workmen climbed the structure and attached themselves to the spokes as the wheel turned.

I'm blown away when I think about engineering marvels like the Ferris Wheel. Just to dream up an idea like that is one thing, but having the confidence to know that it's going to work, when you think about the ridiculous size of the axle, is just incredible. I'd love to have been in that room discussing the maths of it, and then fast-forward to that moment years down the line when their sums are being very seriously tested. But it worked a treat, and became the centrepiece of the fair. Not only that, it got the whole world talking. Around 1.5 million people went for a ride on the Ferris Wheel, paying $0.50 cents each, which should have brought in $750,000, but Ferris and his investors were locked in a litigation fight with the fair's organisers for two years afterwards. And then, tragically, Ferris's life was cut short: he died of typhoid fever in 1896, at the age of just 37. But, his Ferris Wheel lived on. Somehow they managed to dismantle it and rebuild it in Chicago's North Side where it stayed until 1903, before it was transported by rail to St Louis, Missouri, for the 1904 World's Fair. That's right – Ferris's idea was so good, it was used for two World's Fairs. And Ferris's legacy continues, because as you look across any fairground, the first thing you're likely to lay your eyes on is a Ferris Wheel.

FERRIS WHEEL
The sheer scale of George Ferris's wheel was bonkers, but he pulled it off for the World's Fair in 1893 and its legacy has lived on ever since.

HOME AND RECREATION

MACHINE NUMBER	MACHINE NAME	PAGE NUMBER
089	HAND DRILL	204
090	CARPET SWEEPER	206
091	POWERED VACUUM CLEANER	210
092	DISHWASHER	214
093	HYDRAULIC JACK	216
094	HAND-POWERED LAWNMOWER	218
095	PHONOGRAPH	220
096	BICYCLE	222
097	CLOTHES MANGLE	227
098	EXERCISE BIKE	228
099	ROWING MACHINE	230
100	TREADMILL	232

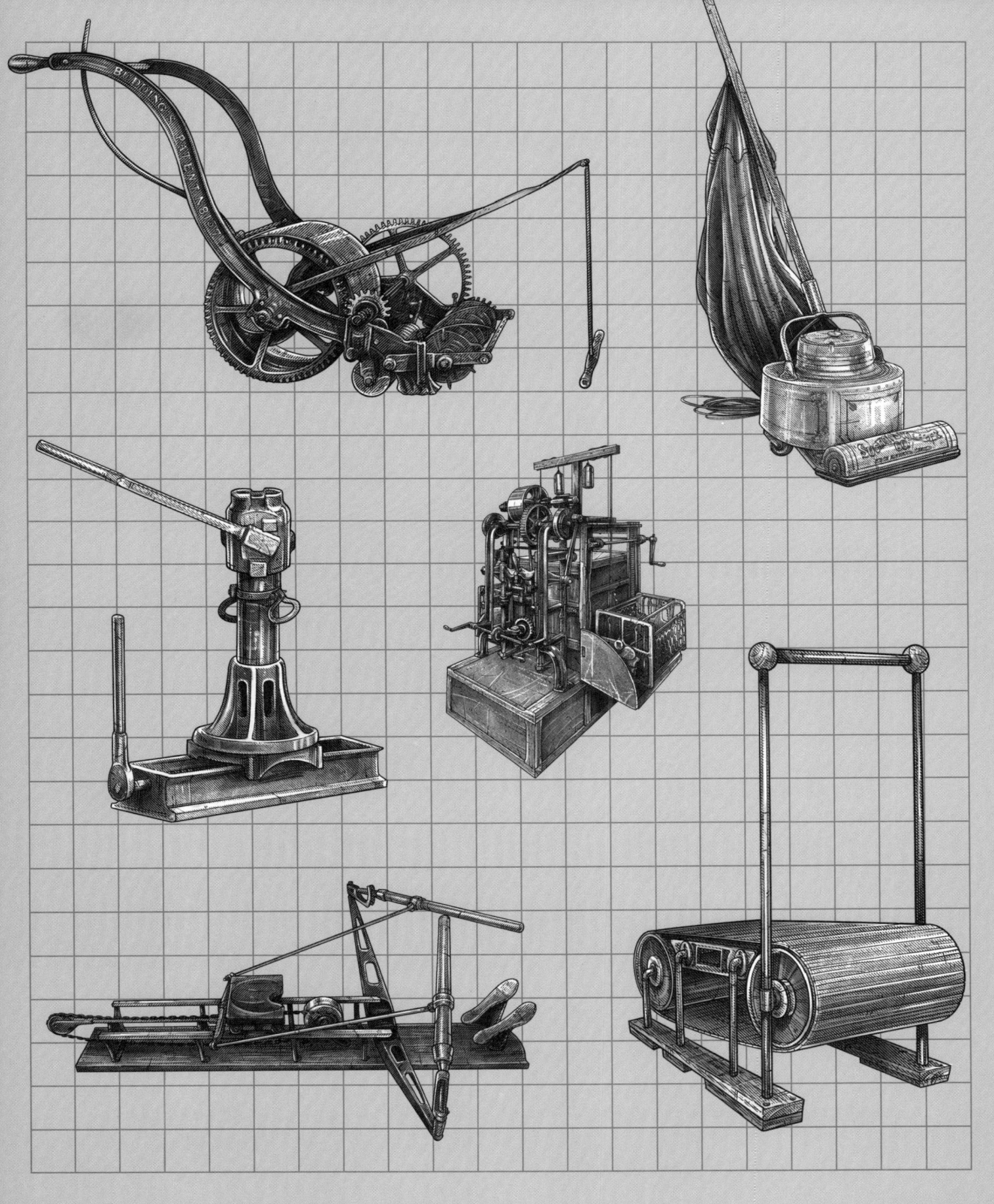

HAND DRILL

MACHINE NUMBER 089

A hand drill is a classic tool that you'll often find in the back of grandad's shed. In the UK, they often bring back memories of woodworking classes at school, where you're tasked with making something like a wooden plant pot. Hand drills have been around for a long time, and they haven't changed very much at all. The first known picture of a geared hand-drill dates back to 1816 in a book called *L'Art du Tourneur*, and it seems likely that they originated in France sometime in the 18th century.

Before the geared drill was developed, the tools people used for drilling – augers, pump drills and bow drills – were really quite primitive. The "hand brace" was an important development in the Middle Ages because it introduced a continuous drilling motion with its U-shaped, spinning design. The earliest brace anyone has found was in the wreck of the *Mary Rose*, King Henry VIII's flagship, which famously sank in the Solent in 1545 and was raised in 1982.

A geared drill, also known as an eggbeater drill for obvious reasons when you look at the illustration, has a hand crank attached to the main gear. When the main gear spins, it turns a horizontal pin gear on a shaft that is connected to a chuck (a specialised small clamp to hold a drill bit in place). Geared drills formed two categories – hand drills and breast drills. Hand drills were a bit smaller (38cm/15in and under) and better for drilling into wood. Breast drills were longer and featured a concave plate at the other end to the drill bit, which the user leant against to transfer as much power as possible. They also freed up the hands of the user to turn the crank. It's for these reasons they were better suited to boring holes in metal and very hard wood. Early geared drills had a gear ratio of 1:1, which meant that the chuck spun once for every complete turn of the drive wheel. But by the 1880s, you'd be seeing geared drills with a gear ratio of over 4:1, so each turn of the drive wheel would make the chuck spin four times as quickly.

Millers Falls, established in Greenfield, Massachusetts, in 1868, was a big name in manufacturing and it produced some of the most respected geared drills around. In its 1915 catalogue, Millers Falls produced 28 different hand drills and 40 breast drills. These could even be mounted in specialised metal frames to form a drill press.

Its No. 13 breast drill, which was first produced in 1881, provided a gear ratio of 4.5:1 thanks to its 15cm/6in-diameter drive gear, and double pinions that afforded it more stability. This made it ideal for drilling into metal, although it came at a hefty cost at the time, which seems to have been something the company was quite proud of in its catalogue! It was a really successful drill and produced all the way until the mid 1930s. Miller Falls' most famous hand drill, though, was probably the No. 2, introduced in the 1870s. Although it was modified over the years, the basic design was still much the same, 110 years later.

The first drill I ever bought was a hand drill, which I picked up at a boot sale. They are great little machines for getting children into woodworking, and the drill bits can be stored in a secret compartment in the handle.

And with a keyed drill chuck, kids won't be able to loosen the bits. I don't really need to use a hand drill that much these days, but Lucia Scalisi, our painting conservator at *The Repair Shop,* does all the time and she swears by them. She uses them to drill pilot holes in the back of stretchers or frames so she can screw on the little brackets and hooks that hold the stretcher in the frame. For drilling small holes in soft wood, which is something that needs to be done slowly and carefully, a hand drill is ideal. You just crank the handle and push down gently. I've helped her out on a few jobs, and it's the scariest thing using an electric drill to bore into the back of a frame knowing there's a valuable painting on the other side which Lucia has spent weeks conserving and restoring. One slip, and you'll go all the way through.

THE FIRST DRILL I EVER BOUGHT WAS A HAND DRILL, WHICH I PICKED UP AT A BOOT SALE.

HAND DRILL
The geared drill was also known as the eggbeater drill because of its appearance. Most hand drills have a gear ratio of 4:1, so for every turn of the handle, the drill bit will turn four times.

CARPET SWEEPER

MACHINE NUMBER 090

The good old broom and dustpan-and-brush combo was fine for hard floors, but a bit of a nightmare on rugs. Cleaning them generally involved picking up the rug (which is always about twice as heavy as it looks), dangling it out of a window and beating it repeatedly with a stick. But when fitted carpets became popular in the 19th century, I imagine most people were soon saying: "That looks great, but how on earth are we going to clean it?!"

Two folks who were suffering more than most were husband and wife Melville and Anna Bissell, whose surname will be familiar to anyone in the market for a carpet cleaner. In 1870, the Bissells opened a shop in Grand Rapids, Michigan, selling crockery and glassware, and the story goes that when the fragile items arrived at their shop, they were always packed in sawdust, which inevitably made its way onto their carpeted shop floor – and that's not really the look you're after unless you're running a butcher's shop. Trying to sweep it up with a dustpan and brush would have tested their patience, and possibly even their marriage, so something had to be done. Melville bought them an early mechanical carpet sweeper, but it wasn't very good, so he started working on his own design.

In 1876, Melville patented his lightweight carpet sweeper, featuring a central bearing brush and rubber tyres. Unlike the competition, it picked up dirt easily, worked on bumpy, uneven floors and was light and easy to use. It was a stroke of genius and suddenly all their friends, family and customers wanted one. So the Bissells started selling their carpet sweepers door-to-door, and these were put together by a handful of employees in the room above their store. A few years later, the Bissells were the market leader, and had built their own five-floor factory to make their hugely popular product. Sadly, in 1884, a massive fire broke out, gutting their factory, but the Bissells sucked it up, rebuilt and expanded their business. But then tragedy struck – Melville came down with pneumonia and died, aged just 45, in 1889. Anna took charge of the company, becoming the first female CEO in the US. During her tenure, Bissell picked up a famous fan, no less than Queen Victoria, and the business went from strength to strength. Anna only stepped down in 1934 at the ripe old age of 87, and the company is still owned by the Bissell family today.

A rival company across the Atlantic began producing its own successful carpet sweeper in 1889. Based in Accrington, Lancashire, Entwistle & Kenyon, which had been producing washing machines and mangles until then, branched out with its Ewbank sweeper, named after an area of Accrington where the factory was situated. Designed by Richard Walter Kenyon, after seeing how popular carpet sweepers had become during a trip to the US in the 1880s, this carpet sweeper became such a hit in the UK that "ewbanking" became a verb. The company still makes carpet sweepers today.

CARPET SWEEPER
With the arrival of fitted carpets in the 19th century, the broom was no longer up to the job of keeping the house clean. The answer was the carpet cleaner.

IT PICKED UP DIRT EASILY, WORKED ON BUMPY, UNEVEN FLOORS, AND WAS LIGHT AND EASY TO USE. IT WAS A STROKE OF GENIUS.

POWERED VACUUM CLEANER

This machine shouldn't really sneak into a book of mostly manually powered, steam-powered, pneumatic and hydraulic machines, because it does feature an internal combustion engine, but I got so involved in the story that I couldn't help it.

The British civil engineer Hubert Cecil Booth spent most of the 1890s working for Maudslay, Sons and Field in Lambeth, south London, a company specialising in manufacturing marine steam engines. It also, famously, made a steam-powered mill in 1852 to re-cut the Koh-i-Noor – one of the most famous diamonds ever found and cut, which remains part of the British Crown Jewels. Booth spent his time at the firm designing bridges, battleship engines and amusement park rides in London, Vienna and Paris, among other places. But everything changed for Booth in 1901 after he went to see an American inventor demonstrating his new machine at the Empire Music Hall in London. The machine was said to blow dust off carpets using jets of compressed air, and could well have been the "pneumatic carpet renovator" designed by John S. Thurman, an inventor from St Louis, Missouri, who'd submitted a patent application for his device in 1898. Booth met the inventor at the event and asked him why the machine didn't suck the dust up rather than blow it. Perhaps unsurprisingly, this wasn't received very well. Here is Booth's recollection of the conversation: "the inventor became heated, remarked that sucking out dust was impossible and that it had been tried over and over again without success; then he walked away."

What Booth lacked in small talk, he made up for in ingenuity. Well, eventually anyway. He navigated a few challenges first, including, apparently, nearly choking to death during an experiment in a restaurant with a handkerchief "filter" over his mouth, to demonstrate that he could suck up dirt from a chair onto the handkerchief. After recovering from that incident, Booth put his money where his mouth was, and set to work designing a machine that sucked. It didn't take him long – the same year he'd seen the

CROWDS OF PEOPLE WATCHED THE DUST AND DIRT BEING SUCKED UP INTO A GLASS RECEPTACLE, AND IT BECAME THE TALK OF THE TOWN.

THE NEW HOOVER

Advertisement for "The New Hoover" – an upright vacuum that brags an easier and quicker clean.

Empire Hall demonstration, he'd both applied for a patent and established the British Vacuum Cleaner Company (BVC). His machine, which was the size of a small van, was horse-drawn, and used a petrol-driven motor to suck air through a hose and a filter. It was nicknamed the "Puffing Billy", and it must have put on quite the circus show, travelling the streets of London's posher neighbourhoods with its red and gold liveried cart and vacuum operators, who fed long rubber hoses up through the windows of the upper floors. Crowds of people watched the dust and dirt being sucked up into a glass receptacle, and it became the talk of the town. There were a few drawbacks. The noise must have been pretty unbearable and Booth faced a number of court cases from rival inventors, but he was getting some high-profile gigs, including King Edward VII's coronation at Westminster Abbey in 1902.

Booth's machine wasn't just an entertaining sideshow, though; dust samples it sucked up were revealed to contain diphtheria bacteria, which led to BVC being awarded the Royal Sanitary Institute's Rogers Field Gold Medal. Suddenly, Booth's device wasn't just celebrated for making places look cleaner; there were also serious health benefits attached, too.

Meanwhile, in the US, the inventor David T. Kenney had designed a similar device, although his first ones were stationary, like the vacuum cleaning machine installed in a building owned by the industrialist Henry Clay Frick in Pittsburgh, in 1902. Kenney was later hailed as "the father of the vacuum cleaner industry" by the *New York Times*.

The next challenge was making the vacuum cleaner smaller, more portable, affordable and suitable for home use. English inventor Walter Griffiths developed one of the first in 1905, patenting it as "Griffith's Improved Vacuum Apparatus for Removing Dust from Carpets", and it does resemble a modern vacuum cleaner up to a point, only you'd use one hand to compress the bellows, and the other to move a nozzle along the floor.

As for the word "Hoover" becoming a synonym for vacuum cleaner in the UK, Ireland and other countries, that story began in 1907 with a down-on-his-luck inventor and salesman called James "Murray" Spangler. Spangler was nearly 60 years old, and working as a janitor at a department store in Canton, Ohio. He was an asthmatic, and sweeping floors in a big store wasn't exactly the ideal occupation for him. So, as with many stories in this book, he designed a machine to help sort out the problem that was annoying him. He modified a manual carpet sweeper, attaching an electric ceiling fan motor to the top of it, and cutting out a section of the back of the sweeper so he could mount fan blades in the hole, which directed dirt into, amazingly, a pillow case he'd asked his wife for. A broomstick served as the handle, and a rotating brush, inspired by a street-sweeping machine, helped gather up the dirt. He later used a tin soapbox as the housing. Put all that together, and you've got an experimental, but early, version of an upright electric vacuum cleaner.

Spangler refined his device, called it the "suction sweeper", applied for a patent and formed the Electric Suction Sweeper Company, with money invested by a friend. But he didn't have the resources to scale up production. He needed more cash and more employees. His cousin's husband, William Hoover, who then owned a leather manufacturing business, saw the potential and bought Spangler's patent in 1908. Spangler was hired as production supervisor by The Hoover Company. His wife and daughter were also employed, making bags for the sweepers. Spangler negotiated a royalty deal until his patent expired in 1925. Sadly, in 1915, he died, aged 66, the night before he was about to take the first holiday of his life, to Florida. We've talked about a lot of unlucky inventors in this book, but this story might just take the biscuit.

The Hoover Company's first vacuum cleaner was the Model O, but it wasn't an immediate success. So, later in 1908, Hoover placed an advertisement in the popular magazine *The Saturday Evening Post* offering potential customers free use of one of its vacuum cleaners to anyone who wrote to them and asked for one. Rather than send the vacuum cleaner directly to the customer, Hoover negotiated with local stores to supply the device, and keep a commission if the customer bought it after the trial period. Not only was this a successful commercial gamble in the US, it also laid the foundations for the company's national dealer system of a network, and stirred international interest in its products.

Hoover Limited became a registered company in the UK in 1919, and opened a huge building in Perivale, London, in 1932, which served as the UK headquarters, production plant and repair centre for the Hoover Company. You can't miss The Hoover Building on the A40 on the way out of London – it looks like an Art Deco palace and is one of the most iconic 20th-century buildings in the country. Designed by Wallis, Gilbert and Partners, the building with its remarkable façade, its distinctive green window frames, and its two elaborately decorated towers featuring Aztec and Mayan motifs, was Grade II* listed in 1980. It has always been dramatically lit up with green floodlights at night, and its railings and scalloped windows feel like something straight out of *The Great Gatsby*.

After the war, Hoover became such a big force in the UK vacuum cleaner market that by the late 1950s, the word became a generic trademark. Although it doesn't have anything like the market share it used to, "hoover" is still the word almost every Brit uses for vacuum cleaner.

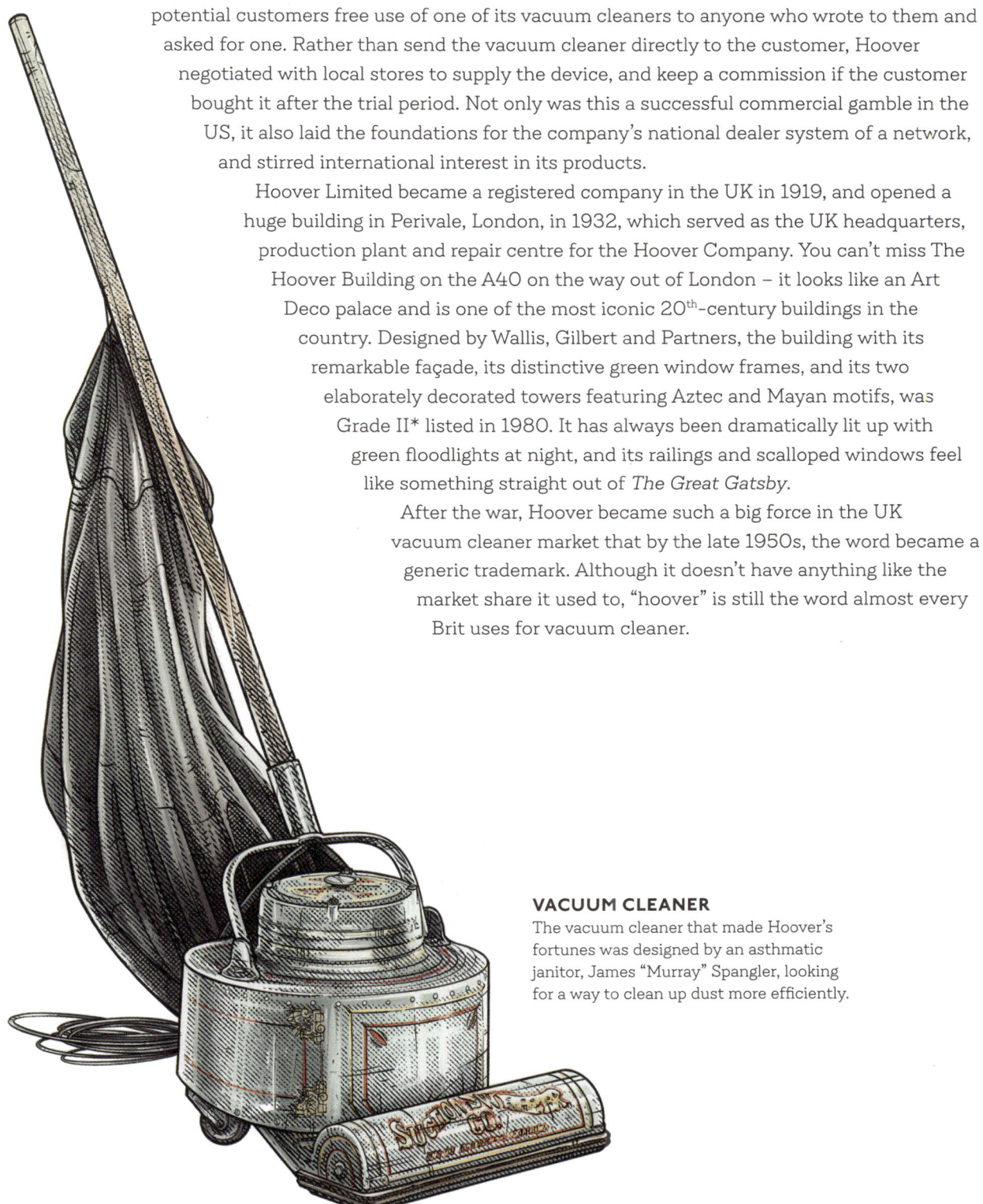

VACUUM CLEANER
The vacuum cleaner that made Hoover's fortunes was designed by an asthmatic janitor, James "Murray" Spangler, looking for a way to clean up dust more efficiently.

214 | CHAPTER 6 | HOME AND RECREATION

DISHWASHER
Early dishwashers were hand-powered, but even though an electric one was launched in Europe in 1929, the device didn't really become popular until the 1970s.

DISHWASHER

MACHINE NUMBER 092

The first evidence of a machine to wash dishes was a patent granted in 1850 to the American inventor Joel Houghton. It was a simple device that splashed water at dirty dishes in a wooden tub, and was operated by a hand-cranked wheel. It didn't take off. But, just over 35 years later, Josephine Cochrane devised something much more sophisticated. Josephine had come from a family of inventors and engineers and she was a natural problem-solver.

Married to a successful merchant and politician, Josephine had become a wealthy socialite, accustomed to holding fancy dinner parties. The trouble was, that their equally fancy, delicate china kept getting chipped when it was hand-washed. No one else seemed to be interested in inventing a mechanical dishwashing machine that could clean effectively and gently, so she got on with it herself. She came up with a copper boiler containing a wooden wheel that housed a number of wire compartments for different items of china to sit in – one for the cups, one for the bowls, etc. Beneath the wheel was a container for the soap. The wheel could be hand-turned, but later models were powered by a motor driving a pulley system. In 1886, she founded the Garis-Cochrane Manufacturing company to produce her dishwashers.

Josephine's invention was another that benefitted from the timing of the 1893 World's Fair, in Chicago, because it gave a much bigger stage to her remarkable invention, which she named Lavadora (and later changed to Lavaplatos, as Lavadora was already in use). Nine of her dishwashing machines were used by the restaurants at the fair, and after seeing them work, she was receiving orders left, right and centre from hotels and restaurants, and anyone with a big kitchen and a ready supply of hot water. Her company was renamed Cochrane's Crescent Washing Machine Company, which later became part of KitchenAid.

Miele launched Europe's first electric top-loading dishwasher in 1929, but the Wall Street Crash the same year (and the Great Depression that followed) meant that people weren't willing to spend a small fortune on something that was perceived to be a luxury. It wasn't until the 1970s, nearly 85 years after Josephine's visionary invention, that dishwashers became a common appliance in North America and western Europe. But having said that, I've never actually owned a dishwasher. I think it has something to do with the fact that I was the pot washer in my parents' restaurant. We had a special glass washer for the glasses, but everything else was washed up by yours truly in the sink. You'd have thought that would make me want a dishwasher, but I prefer to do everything by hand if I can. And it's a good way to clean off the last of whatever is on my hands after a day in the workshop, or at *The Repair Shop*.

HYDRAULIC JACK

It might not surprise you to learn that jacks have always been a part of my life. My brothers and I spent hours on our front driveway messing around with cars and jacks. Saturdays involved heading to scrapyards, and we had an awesome time. It was a whole day out to us. On Sundays, we went to boot sales and over time we must have picked up about half a dozen massive trolley jacks, and most of them were made by the classic British brand Bradbury. It is still around today, making two-post lifts, scissor lifts, and four-post lifts among other automotive equipment.

The jacks we bought back then were completely impractical purchases, because we had nowhere to put them, and they were too heavy to lift, so the wheels would always end up sinking into the mud. Plus, the seals were knackered, so the jack would just gradually sag back down after you'd pumped it up. It turns out that they were in a boot sale for a reason. But we were taught well, so we learned to jack up a car and then quickly put axle stands underneath it, before the jack started sagging. It's an easier and simpler story these days. I don't often buy new equipment but I did purchase a new Snap-On hydraulic jack. It works absolutely perfectly, but part of my brain is still always thinking, "Quick – get the axle stands".

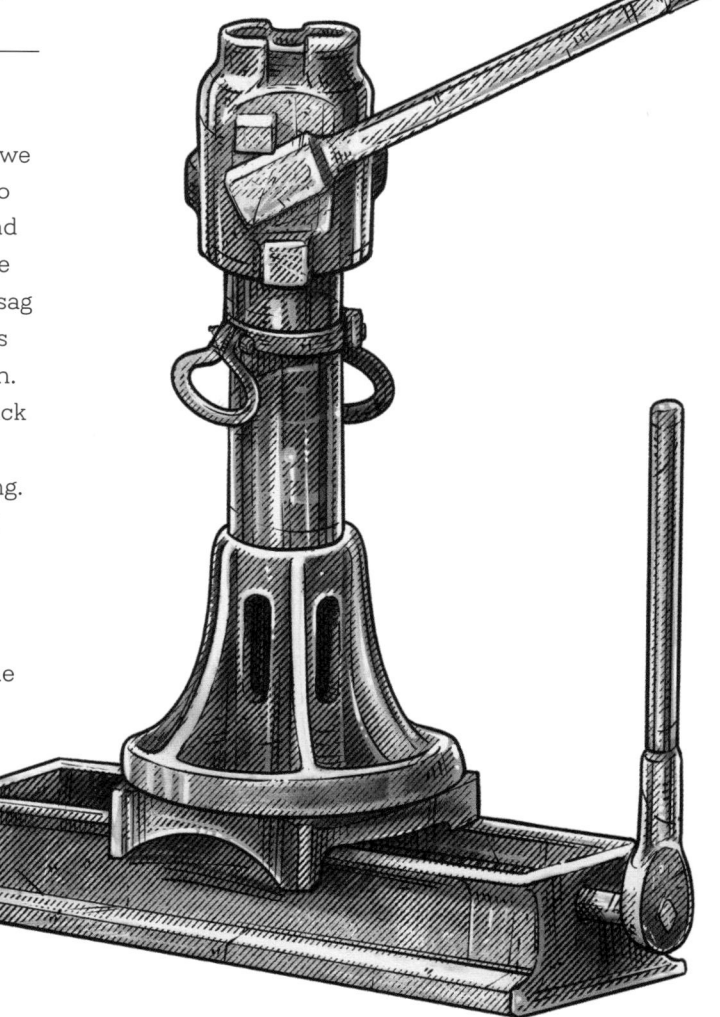

HYDRAULIC JACK
Sometimes the liquid used in early, portable hydraulic jacks was whisky because it froze at a lower temperature than water.

The hydraulic jack goes all the way back to 1838, and an inventor called William Joseph Curtis whose invention could "place or replace an engine or carriage upon the rail". It was an adaptation of the hydraulic press, and involved two upright cylinders "each furnished with a ram or piston", which were mounted on a moveable plank. Between the two cylinders was a pump, operated by a bell-cranked lever. Water was fed into the pump from a small cistern, also mounted on the plank, and valves enabled the water to pass from the cistern to the cylinders. Cranking the pump elevated the rams, which raised the engine or carriage.

In 1852, engineer Richard Dudgeon was granted a patent for his portable hydraulic jack (although he called it a "press" at the time). He was originally from Scotland, but moved to New York with his family when he was young. Working as an apprentice in a machine shop, Dudgeon was used to lifting heavy equipment with screw jacks, which hadn't actually changed that much since the days of Ancient Rome. So he went about devising something portable, and much more efficient. His first jack, built in 1851, was a simple cylinder (roughly 13cm/5in in diameter) with an enlarged hollow head that served as the fluid reservoir to supply the cylinder with "water or other liquid". Amusingly, the other liquid was occasionally whiskey, which froze at a lower temperature than water, which led to the nickname "whiskey jack". Like modern jacks, you just cranked a handle repeatedly, and you'd be able to lift 100 tons of equipment. Hydraulics never cease to amaze me – crank a handle a few dozen times with one hand, and you've lifted something weighing the same as 15 elephants. It's crazy. Dudgeon later changed the location of the reservoir to the base of the jack, most likely to stop the jack from being so top heavy, but it didn't stop his first jack from becoming popular very quickly. Dudgeon jacks were used to lower a Cleopatra's Needle[1] – the nearly 3,500-year-old huge stone obelisk – from its site in Alexandria in Egypt, and then to raise it in its current position in New York's Central Park.

CRANK A HANDLE A FEW DOZEN TIMES WITH ONE HAND, AND YOU'VE LIFTED SOMETHING WEIGHING THE SAME AS 15 ELEPHANTS. IT'S CRAZY.

HAND-POWERED LAWNMOWER

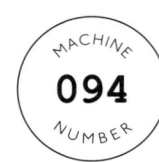

The first mechanical lawnmower was invented by the English mechanic and machinist Edwin Beard Budding, of Stroud, Gloucestershire, in 1830. He was said to have been inspired by observing a machine in a cloth mill which used rotary blades to trim any bobbles off woollen cloth to give a smooth result. So he wondered if he could apply the same principle to cut grass. Until that point, grass was typically cut with a scythe, and if you've ever tried this, you'll know that it's back-breaking work, takes ages, and produces a pretty crap result.

Budding was granted a British patent in August 1830 for "a new combination and application of machinery for the purpose of cropping or shearing the vegetable surface of lawns, grass-plats and pleasure grounds". He went into partnership with John Ferrabee, who owned the lease to a local mill, and had built a foundry there. The first lawnmower they produced had a 48cm/19in-wide frame made of wrought iron, and featured a large roller in contact with the ground, which drove the gears that turned the blades on the cutting cylinder. The whole contraption was pushed from behind like a modern lawnmower, but the grass came out of the front of the machine and was collected in a tray. It most probably required serious effort to push it, given what Budding stated in his patent application: "Country gentlemen may find in using my machine themselves an amusing, useful and healthy exercise" (which sounds like 19th-century code for bloody hard work). Also, a handle was later added at the front to help draw it forwards from the front, making it a two-person job. Still, it was far better than anything that had been invented up to this point, and it became popular quickly. And this is clear because one of the earliest Budding lawnmowers was sold to the Regent's Park Zoological Gardens in London.

Ferrabee didn't have the means to produce machines in the kind of quantities needed, though, so J. R. & A. Ransome of Ipswich, a well-known producer of agricultural ploughs, was granted a licence to manufacture Budding's design. By 1840, over 1,000 Budding lawnmowers had been sold.

I've fixed two lawnmowers on *The Repair Shop*, and the first one was a much-loved unpowered Ransome's model, made in the mid to late 1940s. It was owned by a chap called John, and his grandad had maintained it with all the love and attention that you would for a classic car. You could make out the coat of arms of King George VI proudly displayed on the grass box, because the company had been awarded a prestigious royal warrant. The second mower I fixed also had a royal connection, because I was working on it when we had a visit at the barn from His Royal Highness The Prince of Wales (as he was then, now King Charles III, of course). Jay Blades introduced me to him while I was taking apart a 1950s mower made by Atco, another famous British brand. It turns out that King Charles really knows his lawnmowers: he'd actually awarded the company a royal warrant in 1986. He told me that he'd been trying to convert a petrol-driven mower to run on vegetable oil. We had a cup of tea and got chatting about bees – we're both keen beekeepers.

He asked if we had any honey for his Earl Grey tea and as luck would have it, I had a pot of my homemade honey at the barn. What struck me was how passionate and knowledgeable he was in general, but specifically about heritage craft and apprenticeships. It was a real privilege to get to spend some time with him.

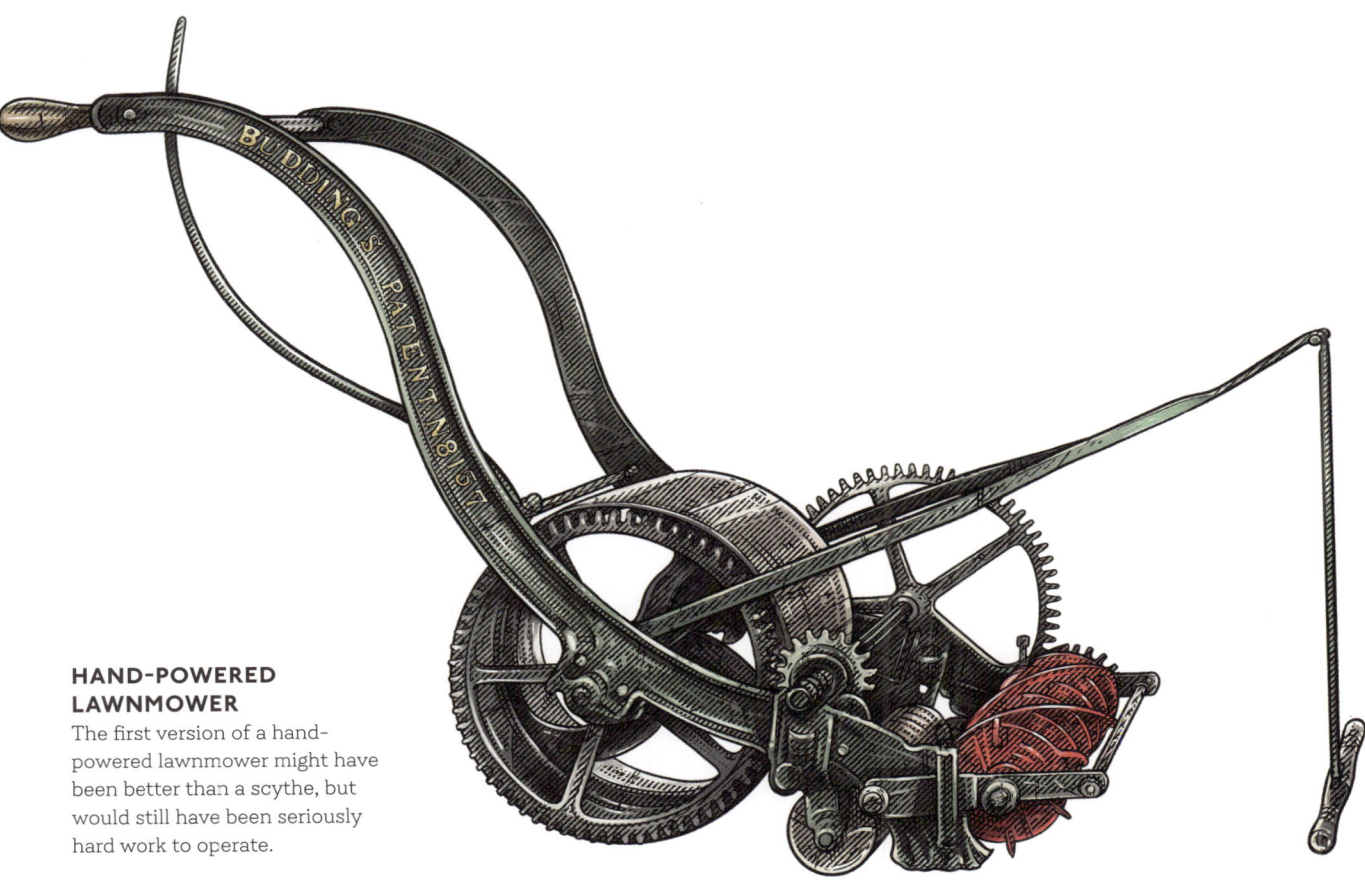

HAND-POWERED LAWNMOWER
The first version of a hand-powered lawnmower might have been better than a scythe, but would still have been seriously hard work to operate.

IT TURNS OUT THAT KING CHARLES REALLY KNOWS HIS LAWNMOWERS: HE'D ACTUALLY AWARDED THE COMPANY A ROYAL WARRANT IN 1986.

PHONOGRAPH

The story of the phonograph involves some of the most famous inventors of the late 19th century competing head-to-head. Although from the 1850s many folks had been working on experimental versions of a mechanical instrument to record and reproduce sounds, it was 30-year-old telegrapher turned inventor, and entrepreneur, Thomas Edison who came up with the phonograph in 1877.

It worked like this: a person talks into a funnel, which "collects" the sound waves, amplifies them and directs them towards a "diaphragm" – a thin membrane, much like the skin of a drum. The diaphragm is connected to a stylus (basically a metal pin) and a cylinder wrapped in tin foil. The sound vibrations make the stylus move up and down, forming a groove in the foil. While this is all happening, someone needs to crank a handle to keep the cylinder moving very slowly, so you end up with a continuous groove. And there you have it – a recorded version of the sound you've just heard. To hear the sound back, you run the stylus back over the groove, which sends the original vibrations onto the diaphragm, and out through the funnel. Edison's first public demonstration of his phonograph is described as follows: "In December 1877, a young man came into the office of *Scientific American*, and placed before the editors a small, simple machine about which very few preliminary remarks were offered. The visitor, without any ceremony whatever, turned the crank, and to the

PHONOGRAPH
Although Thomas Edison and Alexander Graham Bell were behind the early invention of the phonograph, it was a third-player, Emile Berliner, who made the crucial developments that ensured its success.

astonishment of all present the machine said: 'Good morning. How do you do? How do you like the phonograph?'"

Although the folks in the *Scientific American* office were amazed, they basically saw it as a scientific curiosity or an amusing toy. The next step for Edison was improving the quality of the recording, finding a different, longer-lasting substance instead of tinfoil for the recording medium, and devising a way to reproduce an existing recording. And that's when Edison came up against fierce competition, in the shape of Alexander Graham Bell, the man credited with inventing the telephone, in 1875. Bell set up the Volta Laboratory in Georgetown, Washington D.C., partly to explore how to help deaf people communicate, but also to improve Edison's phonograph. By 1885, Bell, along with his cousin Chichester A. Bell, and instrument-maker and inventor Charles Sumner Tainter, had created the graphophone, an improved version of the phonograph which recorded the sound onto a wax-coated cardboard cylinder, and cut a groove vertically rather than horizontally. This all helped to capture longer recordings and increase the quality of the audio when playing the sound back. Edison, meanwhile, had improved his machine in much the same way, and both sides had applied for patents. Edison had even begun to market pre-recorded pieces of popular music on wax cylinders. But then, someone else entered what had looked like a two-horse race, and this guy had some serious pedigree. The German-American inventor Emile Berliner was known to Bell – Berliner had sold him the patent to his first major invention, the carbon microphone, which became a component of Bell's telephone and made Berliner a very wealthy man.

Berliner had been studying the benefits and drawbacks of both the phonograph and the graphophone, and he came up with some major improvements before anyone else. His experimental machine created lateral incisions, caused by the stylus vibrating from side to side onto a flat disc. The flat discs became known as records, and they could be played on Berliner's reproducing machine – the gramophone. Berliner later created a metal reverse (a negative) of the disc as a mould to make zinc copies from. He was awarded a patent for his gramophone in November 1887, and by July 1889 he was using vulcanised rubber for his disc copies (which was replaced with shellac from 1896). In August 1889, he started selling his discs commercially in Europe. By 1891, Berliner was using a spring mechanism to power his gramophones rather than the hand crank. In 1893, he established the United States Gramophone Company and opened a factory and showroom in Baltimore in 1894, producing standardised 7-inch records. Berliner sold over 700,000 of his discs in 1898 alone.

ALTHOUGH THE FOLKS IN THE *SCIENTIFIC AMERICAN* OFFICE WERE AMAZED, THEY BASICALLY SAW IT AS A SCIENTIFIC CURIOSITY OR AN AMUSING TOY.

BICYCLE

In 1817, Karl Drais, an aristocratic German inventor, developed a two-wheeled machine that the user propelled by pushing their feet along the ground.

MACHINE NUMBER 096

It didn't have pedals or brakes and was referred to by Drais as a *laufmaschine* (running machine), although the French words *draisienne* and *vélocipède* became more widely used terms. It was a balance bike really and it must have been utterly horrible to ride, given that it was made almost entirely of wood. Add that to the truly terrible conditions of the roads, and it must have felt like you were being beaten up for half an hour, but it was a major milestone in the history of the bicycle. The London coachmaker Denis Johnson bought one of Drais's *draisiennes* and set to work trying to improve it. His modified version, which he managed to patent in 1818 was known as a "pedestrian curricle". He produced over 300 of these contraptions, which became widely known as "hobby horses". They were so expensive that only aristocrats could afford them, which is why they were nicknamed "dandy horses" – because the only people riding them were fashionable dandies.

The next chapter seems to start 45 years later, when the French mechanic Pierre Lallement developed a pedal-driven *vélocipède* (which had become the generic term for these early bicycles) and demonstrated it in Paris in 1863. Lallement travelled to the US and was granted a patent, later selling it to the American entrepreneur Calvin Witty for $2,000.

BICYCLE

Originally, the style of bicycle that we recognise today was called the safety bicycle, because the user could reach the ground with both feet – unlike on a penny farthing, or "ordinary bicycle" as it was originally known.

But oddly, history often seems to overlook the achievements of Kirkpatrick Macmillan, a blacksmith from Dumfriesshire, Scotland. In the early 1820s, he saw a hobby horse being ridden and wondered if he could make a better one. And improvement number one was finding a way to stop the rider's feet making contact with the ground when it was in motion. So, sometime around 1840, he worked on a machine with pedals that were connected to cranks on the (much larger) back wheel of his machine via connecting rods. He is said to have regularly ridden it along the rough country roads to nearby Dumfries – a distance of 22.5km (14 miles) – and it took him under an hour. That's seriously good going. The story goes that Macmillan decided to crank it up a notch and take his hobby horse to Glasgow, nearly 113km (70 miles) away. Aside from inflicting a minor injury to a minor, for which he received a five-shilling fine from the local police, he made it there in one piece, in two days. But like many inventors before him who have found themselves relegated to a footnote, Macmillan wasn't interested in the commercial potential of his vehicle. His hobby horse was exactly that – a hobby. So, without a patent, other inventors copied and adapted his machine, and started selling it.

In 1864, the Parisian inventor Pierre Michaux, together with his son Ernest, were working on a pedal-driven vélocipède with pedals and cranks connected to the front wheel. Unlike Macmillan, though, they had commercial instincts, and connections, and the timing worked out perfectly for them. Two of their customers, brothers René and Aimé Olivier, took their vélocipèdes on an 800km (500 mile) adventure from Paris to Marseille, in the summer of 1865. After experiencing the invention in action, the duo stumped up 50,000 francs for a nearly 70 per cent stake in Michaux's company – much like a modern deal on *Dragon's Den*. That gave Michaux both the means to ramp up production and to develop more efficient models, including their diagonal-framed wrought-iron vélocipède, which became the technological benchmark at the time. And then came along the Exposition Universelle in 1867 – a World's Fair held in Paris which drew around 15 million visitors from all corners of the globe. It was the perfect opportunity to showcase their machine across the streets of Paris. The one drawback was that though many improvements had been made to the vélocipède, comfort wasn't one of them, which is why the nickname "boneshaker" came about around this time. Another nickname also emerged, this one in 1868, and this would be the one that stuck: bicycle.

The late 1860s and early 1870s was a period of major innovations on the bike front, partly because over 100 different French companies were all competing feverishly. This led to the creation of the solid rubber tyre, patented by the French inventor and engineer Clément Ader in 1868, and the wire-spoked tension wheel, which was patented by mechanic Eugène Meyer, also in 1868 – this tension wheel was specifically for bicycles, as the concept of the wire wheel had actually been around since the 1820s. Britain and the US got in on the act, and the epicentre of bicycle production was soon to shift from Paris to… Coventry. The was also the era of the high-wheel bicycle, which enabled both higher speeds – because one complete rotation of the pedals propelled you the entire circumference of the wheel – and better shock absorption. The inventor James Starley, who would later develop the differential gear and improve the chain drive, produced the first successful high-wheel bicycle, called the Ariel, in 1871. The high-wheel bicycle became known as the "ordinary" bicycle during the 1880s to distinguish it from a new type of bicycle: the safety bicycle. It wasn't until the 1890s, when they were nearly redundant, that high-wheelers gained the nickname "penny farthings". That nickname came about because of the relative sizes of both of the wheels – the much larger wheel at the front looked more like the size of a farthing coin (¼ penny)

BICYCLE
Singer bicycles provided both "ladies'" and "gents'" versions, advertised here in 1901.

SINGER CYCLES

Highest Quality Only.

Lists Free on Application.

(LADIES)

Singer Patent Free Wheels.
Singer Patent Locking Bolt.
Singer Patent Brakes, Etc.

THE NEW CENTURY SENSATION.

Perfect Combustion.
No Smell.
Easy to Ride.
Safe.
Reliable.
Simple.

Full particulars and Lists of Gent.'s and Ladies' Motor Bicycles and Tricycles, free on application.

Sole Makers: **SINGER CYCLE CO., LTD.,**
LONDON: 17 HOLBORN VIADUCT.

compared to a much smaller penny. Although they have become iconic machines, penny farthings weren't exactly user-friendly. You had to be especially athletic just to get on the seat, let alone ride the thing. They were also unbelievably dangerous when you consider the fact that many of the earlier models had no brakes, and would often fling riders over the handlebars when they struck objects on the street below. Still, they look great.

The safety bicycle put both of the rider's feet within reach of the ground, and the pedals powered the back wheel so the rider's feet weren't in danger of striking the front wheel. Everything changed when the chain drive (first developed in the late 1860s) was fitted onto a safety bicycle. Before the chain drive, your top speed was effectively limited by the diameter of the wheel with the pedals attached. The chain drive allowed the bicycle to use that incredible concept: mechanical advantage. Each revolution of the pedal turned a large gear, attached by the chain to a small gear. And with half as many teeth on the smaller gear compared to the larger gear, it could take you twice as far – as long as you were on flat ground. Dealing with hills would come later.

The most successful of the safety bicycles was the *Rover*, first produced in 1885 by Starley & Sutton Co of Coventry. With its chain drive (which meant that both wheels could be the same size) steerable front wheel, and tangentially spoked wheel (with the spokes attached to the hub at a tangent, which helps transmit torque from the hub to the tyres), it shares the key features of a modern bicycle. The company was co-founded by John Kemp Starley (nephew of James) and William Sutton in 1878, becoming the Rover Cycle Company Limited in the late 1890s. And in case you're wondering, yes – this was the origin of the car manufacturer Rover.

As for what was going on across the pond, the American Civil War veteran Albert Augustus Pope used $900 that he'd saved up from his military salary and invested in a shoemaker's supply company based in Boston, Massachusetts. In under a year it became the largest business of its kind in the country and turned Albert into a wealthy man. But the turning point in his life came when he attended the 1876 Centennial Exhibition in Philadelphia, and saw a display of James Starley's Ariel ordinary bicycles. Initially, he began importing bicycles from England, but soon bought the American rights to the patents, and began producing them under the trade name Columbia. By the middle of the 1890s, bicycles had become hugely popular in the US, largely thanks to the development of the pneumatic tyre by the Scottish inventor and veterinary surgeon John Boyd Dunlop. The US overtook Britain as the bicycle-producing epicentre of the world, with over 1 million manufactured in 1899 alone.

Bikes have always been a big feature of my life, from my BMX obsession as a kid to buying a fixed-wheel bike to commute to and from home when I was living in London and working in set design. I even copper-plated the steel frame of that fixed-wheel bike. It looked awesome!

BRITAIN AND THE US GOT IN ON THE ACT, AND THE EPICENTRE OF BICYCLE PRODUCTION WAS SOON TO SHIFT FROM PARIS TO... COVENTRY.

CLOTHES MANGLE

MACHINE NUMBER 097

Known as a mangle in the UK, and a wringer in the US, the mangle is a mechanically operated clothes press that squeezes clothes between two large rollers, which are turned with a crank handle or wheel, to force out liquid before they are hung up to dry. The distance between the two rollers could be adjusted by a screw (or tow) mounted on the top of the frame.

Laundry was a very different process in the 19th century. Clothes were boiled in soapy water, rinsed and wrung, before they went into a clothes mangle, and were then dried, starched and ironed. It could take days. And that was just for simple items.

A mangle was usually made from cast iron, and often mounted on wheels so it could be rolled inside or outside. You'd also find "mangle houses" that would press washing for a penny or two. Sometimes, people with a mangle offered this service informally to their neighbours. One such operation was run by Robert and Betty Tasker of Accrington, Lancashire, and this was more special than most because Robert was a master blacksmith, and is believed to have developed the first, geared clothes-mangle around 1850. This might have something to do with the fact that Robert and Betty had 13 children, so the laundry pile would likely have looked Himalayan. Robert didn't ever patent his machine, though, declaring that: "God gives men brains to help his brother, not to line his pockets."

EXERCISE BIKE

MACHINE NUMBER 098

The stationary "bicycle" goes back to the turn of the 19th century (although it wasn't called this at the time, as the word "bicycle" only came into being in 1847). One such example, patented in 1796, was the Gymnasticon, which was a pretty impressive machine despite sounding like one of the bad guys from the *Transformers*.

It was designed in London by Francis Lowndes, a specialist in "medical electricity", to exercise a person's joints and muscles. The user either stood or sat on a wooden seat within a large frame, and could operate a higher flywheel via a crank with their hands, and a lower flywheel with foot treadles. It was geared towards those recovering from physical injuries and the disabled, and could also be operated by a person standing outside the frame, turning an external crank to move the flywheels and treadles.

Fast-forward to the 20th century, and stationary bicycles were still being used for "medico-mechanical" purposes. There were two exercise bikes in the *Titanic's* gym, and the magazine *Cycling* tells us a bit about them in its issue from 25 April 1912, ten days after the sinking: "The huge physical culture department included two home trainers of the Hutchins-Hamilton type, so popular in cycling circles. The pattern is that with the large dial facing the pedaller, a pointer revolving clockwise around the edge, on which are marked figures denoting the miles covered."

Another famous manufacturer of exercise bikes in the early 20th century was Spencer, Heath & George Limited, based in London, England. Its models from the 1920s featured cast-iron flywheels that could be adjusted with a canvas tensioning belt. In front of the handlebars was a wooden disk (with a large arrow for a dial) that rotated slowly as you pedalled – each full rotation of the arrow indicated that you'd travelled 400m (¼ mile). The flywheel was connected to a sprocket that turned a gear, which turned a bar with a small wheel connected via a rope to the large wooden disk. The whole contraption was mounted on a long, rectangular pine board, with three joists that rested on the floor. They seem to go for a fortune when they come up on auction sites.

EXERCISE BIKE
Stationary exercise machines existed long before the word "bicycle". As early as the 1920s they had a dial so you could see how far you'd pedalled.

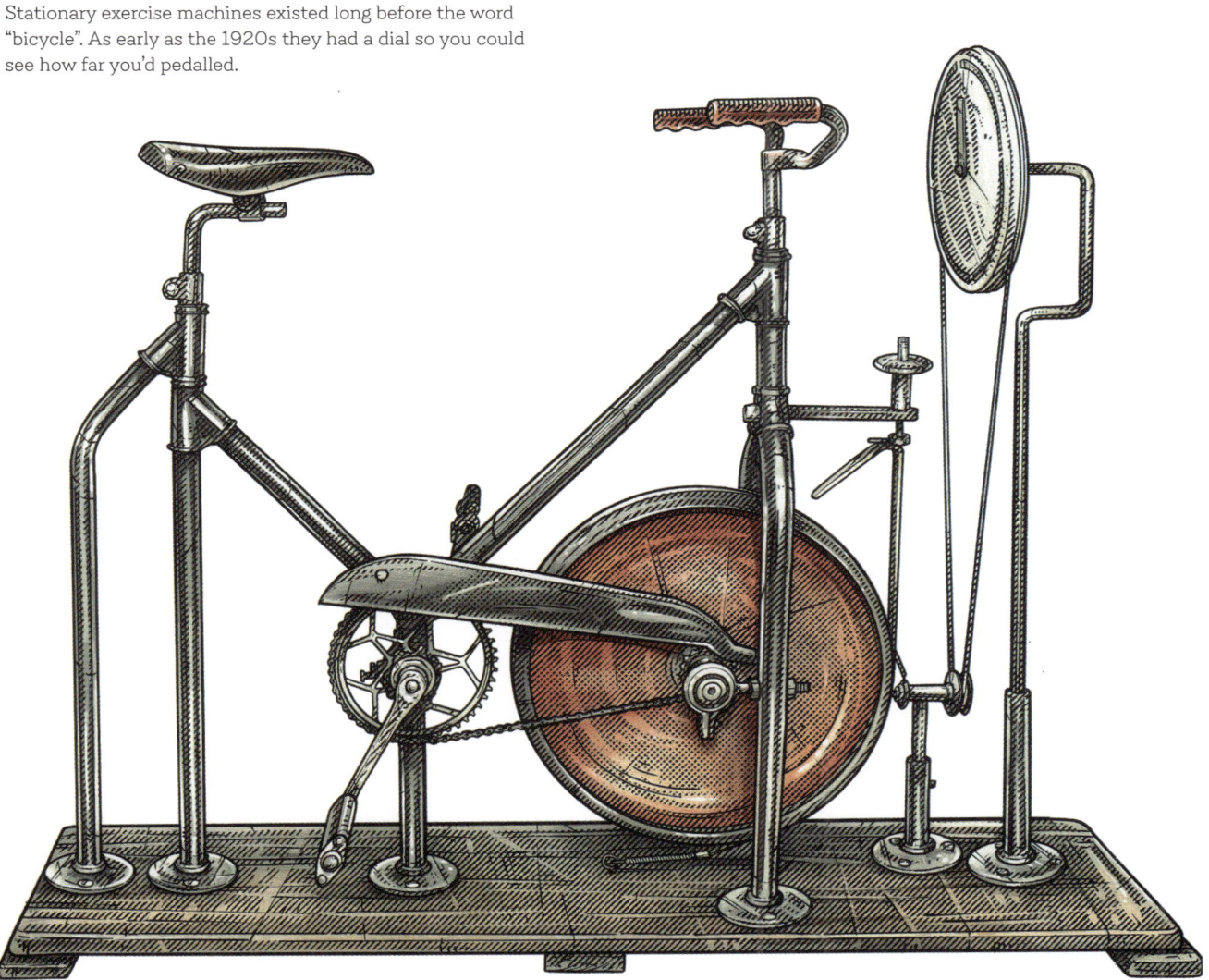

STATIONARY BICYCLES WERE STILL BEING USED FOR MEDICO-MECHANICAL PURPOSES. THERE WERE TWO EXERCISE BIKES IN THE *TITANIC'S* GYM.

ROWING MACHINE

MACHINE NUMBER 099

Rowing machines, or indoor rowers, actually go back to 4 BCE in Ancient Greece. Wooden frames with oars were built on land, so oarsmen could practise their rowing technique for when they were deployed on naval vessels.

In 1871, the well-known US athlete William Buckingham Curtis, who was a three-time national champion in the hammer throw and went on to help found the Amateur Athletic Union (which later became the US Olympic Committee), was granted a patent for his rowing machine. With its flywheel and ratchet system, it's not a million miles away from modern machines.

A.G. Spalding (founded by American baseball player and later executive Albert Spalding, in 1876) got in on the act towards the end of the century, with some beautiful cast-iron rowing machines, which provided resistance thanks to a chain mounted around cogs.

There's a famous photograph of the gymnasium for the use of first-class ticket-holders aboard the *Titanic* in 1912. Apparently, ladies were permitted to use the gym from 9 a.m. to midday each day, and gentlemen from 2 p.m. to 6 p.m. In the centre of the picture is a rowing machine being used by a moustachioed man wearing white flannels. He's Thomas McCawley, the ship's "physical educator" and he's demonstrating the Narragansett hydraulic rower, which was produced between 1900 and 1960 in Rhode Island, US, and which used pressurised gas to create resistance. But like Curtis's machine, it had no way to measure the output of the rower. That changed in the 1950s with the work of the Australian academic Frank Cotton, professor of physiology at the University of Sydney. He developed a machine he called the ergometer, which could identify the potential of competitive athletes. One such athlete that Cotton identified was John Harrison, a gifted Australian oarsman. He had a run of bad luck, though – coming down with appendicitis before the 1952 Olympics and again falling ill before Olympic selection in 1956. The following year, Harrison developed his own rowing machine, featuring a sliding seat and a footstretcher attached to a horizontal flywheel, a "rev" counter and a stroke-rate monitor to measure power output (with an accuracy range of under 1 per cent). It was an incredible innovation for its time but was eventually superseded by the air resistance rowers developed in the 1980s.

ROWING MACHINE
Rowing machines from the 19th and early 20th centuries counted on flywheels and ratchets, cogs and hydraulics for resistance. Today's machines tend to use air resistance.

APPARENTLY, LADIES WERE PERMITTED TO USE THE GYM FROM 9 A.M. TO MIDDAY EACH DAY, AND GENTLEMEN FROM 2 P.M. TO 6 P.M.

TREADMILL

MACHINE NUMBER 100

The treadmill has a surprisingly dark past. The precursor to the modern treadmill was invented by the British civil engineer William Cubitt. Cubitt was best-known as the chief engineer on the Crystal Palace, built to house the Great Exhibition in 1851. Previously, though, in 1817, he had come up with an idea for a new machine, inspired by the sight of convicts sitting idle in prison cells. He developed a machine with the hope that it might reform offenders sentenced to hard labour "by teaching them habits of industry".

Cubitt's "treadwheel" was essentially a really long wooden cylinder that looked a bit like an elongated waterwheel. The exterior of the wheel featured around 40 evenly spaced ledges, and the largest treadwheels could accommodate up to 24 prisoners. They stood side by side, sometimes separated by individual wooden partitions, and would keep climbing upwards to drive the wheel forward. Some of them, like the one in south London's Brixton prison, were connected to gears and heavy stones to grind corn or to pump water. But most didn't operate any other machinery – they were designed as a punishment, as a way of making prisoners atone for their sins, and as a deterrent from committing further crimes. It must have been a pretty horrible experience, both knackering and brain-achingly boring, especially given that prisoners spent up to ten hours a day on the treadwheel, with short rest periods allowed each hour. Ten hours on one of Cubitt's treadwheels was the equivalent of climbing around 5,182m (17,000ft) – just to put that in perspective, it's higher than the tallest mountain in Europe, Mont Blanc. They weren't just tiring and boring, though – there were also serious accidents and fatalities on prison treadwheels.

These treadwheels weren't just used in the UK – four were installed in prisons and psychiatric hospitals in the US, in the 1820s, but mercifully the experiment didn't last long. Meanwhile, in the UK, there were 39 in use by 1895, but they were outlawed after the Prison Act of 1898. One of the most famous inmates forced onto a treadwheel was Oscar Wilde, who'd been sentenced to hard labour at Pentonville Prison in London for "gross indecency".

Anyway, on to the more modern form of punishment – the treadmill. In 1913, New York City inventor Claude Lauraine Hagen was granted a patent for his "training machine", which featured a treadmill belt (made of fabric or rubber) mounted on a rectangular frame with side rails that could be adjusted for the height of the user. The belt could also be inclined up or down and he'd made attempts to reduce the noise it made. It looks very much like a modern treadmill from the patent drawing, but there's no evidence it was made commercially. Manually operated treadmills were being produced in the late 1910s and 1920s and were mainly used in scientific experiments to measure a person's oxygen levels. There were treadmills designed for recreational fitness, but they were pretty basic devices – usually a belt of thin wooden slats mounted around two wheels with two poles to hold onto. You wouldn't be able to do anything more than medium-paced walking on one. By the middle of the 1930s, they were a lot more sophisticated, with a durable fabric belt, an iron frame, and bars to cling on to.

In my set designing days, I'd hire treadmills from special-effects companies and we'd use them to make it look like people were running in various landscapes. So my job was to hide the treadmill by covering it in gravel or grass to make it look like it's a road or a field. All part of the trickery of film.

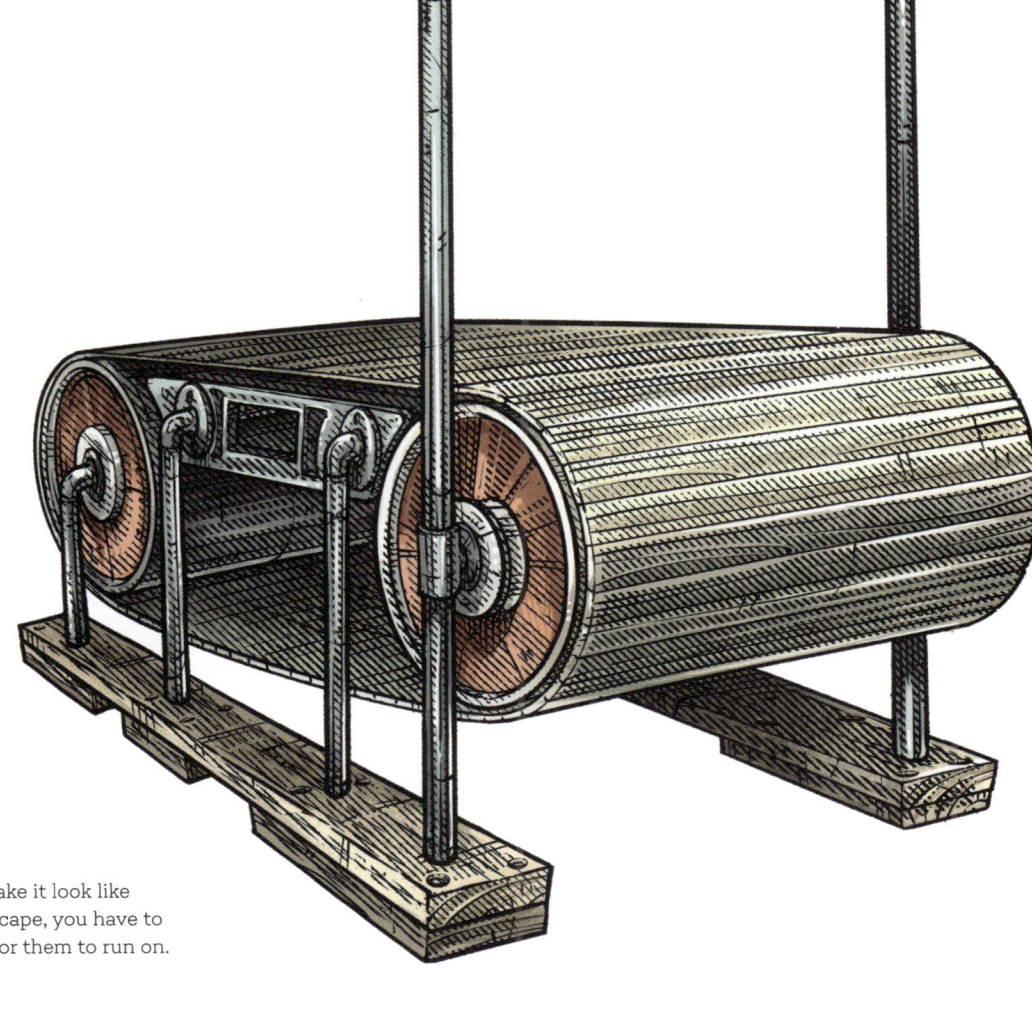

TREADMILL
On film sets, if you want to make it look like someone is running in a landscape, you have to hide and disguise a treadmill for them to run on.

INDEX

bold indicates images

1812 Boulton & Watt engine 133
42 Line Bible 28, 29

A

A1 Heeley Double Lever corkscrew 56, **56**
A.G. Spalding 230
Abraham, David 152
"accident cranes" 26
Adana 85C 100
Ader, Clément 224
Aeolipile 132
Albert, Prince 99, 174
All-American Soap Box Derby Race 194
American Civil War 73, 80, 90, 226
Ames, Nathan 150
Antikythera mechanism 18
Appert, Nicholas 72
Archer Company of St Louis 142
Archimedes 17
Archimedes' screw 15, 19, 140
Archytas of Tarentum 15
Arkwright, Richard 22
Aspinall, Donald 100
Aston Martin 107–08
Atlas 150 Machine 62, **63**
Atlas Pepper Mill 65
Austin J40 189–90
Austin Twenty Ranelagh 108
Ayers, Frederick 100

B

Babbage, Charles 38
bagatelle 180
Baker, William Burton 56
Barham, George 50
Beale, Benjamin 174
Bell, Alexander Graham 58, 219
Belmont 143, **144**
Berliner, Emile 219
Berninghaus, Eugene 142
"Besco" 112
Bissell, Melville and Anna 204
"Blind Jack" 40–41
block and tackle 17
Bonaparte, Napoleon 43, 72, 170
Boucher, C.A. 155
Boulton, Matthew 132
Bourne, William 42
Brachhausen, Gustav Adolf 184
Bradley, Milton 84
Bradshaw, Thomas 182
Bramah, Joseph 92
Bramah Locks Company 92
Bramah water closet 92
Bregeut, Abraham-Louis 128
brilliant cutting 102
British Vacuum Cleaner Company (BVC) 212
Blücher 137
Booth, Hubert Cecil 210
Brown, Joseph 90
Brunel, Isambard Kingdom 127, 140
Budding, Edwin Beard 218
Bundy, Harlow and Willard LeGrand 158
Bundy Time Recorders 158
Bunker, Charles Arthur 73
Bushnell, David 42, 140

C

Cahill, Thaddeus 83
Cambridge roller 159
Carlile, Robert 190
Carters Steam Fair 176, 182
Casler, Herman 184
Caslon Limited 100
Centennial Exposition 1876 58, 226
Charles W. F. Dare Company 182
Chicken Joe 176
Clark, Edward 89
Class 500 146
Clement, Joseph 38
Clephane, James O. 82
Cochrane, Josephine 215
Collier, Ralph 57
Colvin, Lee 154
Cooper, Richard Burton 155
"complex" machine 30
Cooke, William 126
Cooper, Daniel 158
Corbett, Thomas 61
Cotton, Frank 230
Crawshay, Richard 136
Cretors, Charles 173
Cubitt, William 232
Curtis, William Buckingham 230
Curtis, William Joseph 217
Cyclone Seeder Company 164

D

da Vinci, Leonardo 70, 78, 79, 192
Darymple, John I. 104, 108
Davey, Sir Humphrey 137
de Bergue, Charles 95
de Vaucanson, Jacques 79
Deere, John 27
Deleuze and Dutillet 66
Densmore, James 82
des Estivaux, Thierry 155–56
Dickson, William K. L. 184
"Difference Engine No. 1" 38

"diving boat" 42
Dover Stamping Company 57
"Dovers" 57
Drake, Edwin "Colonel" 160
Drais, Karl 58, 222
Drebbel, Cornelius 42
Dudgeon, Richard 217
Durand, Peter 72–73
Durenne, Antoine 94

E
Eckstein, George Frederick 155
Edison, Thomas 118, 163, 184, 220–21
Edmondson, Thomas 156–57
"Eight-Five", the 100
Elizabeth I, Queen 128
Ellis, Peter 148
"Endless Conveyor or Elevator" 150
Enterprise Manufacturing Company 58, 116
Erado 130
Ericsson, John 140
Ewbank sweeper 206
Exhibit Supply Company 176
Exposition Universelle 198, 224

F
F. J. Edwards 104, 112
Farrell, Gabriel 152
Ferguson, Harry 27
Ferrabee, John 218
Ferris, George Washington Gale 198, 199
"fiddle drill" 164, **165**
Field, Cyrus 127
Fisher, John 87
Foecke, Tim 94
Foljambe, Joseph 27
Ford, Henry 163
Ford Model T 163
Ford Motor Company 163
Foucault, Léon 124
Fulton, Robert 42–43

G
Galilei, Galileo 36
Garforth, Mr 98
Gatter Novelty Company 176
Geordie lamp 137
George Salters Spring Company 70
Gibbs, James 90
Girard-Perregaux 128–29
Glidden, Carlos 80, 82
Gottlieb, David 180

grandfather clock 36, 37
Great Depression 129, 142, 194, 215
Great Exhibition 1851 84, 99, 232
Green, George F. 131
Griffiths, E. P. 57
Griffiths, Walter 212
gun worm 54
Gutenberg, Johannes 28–29

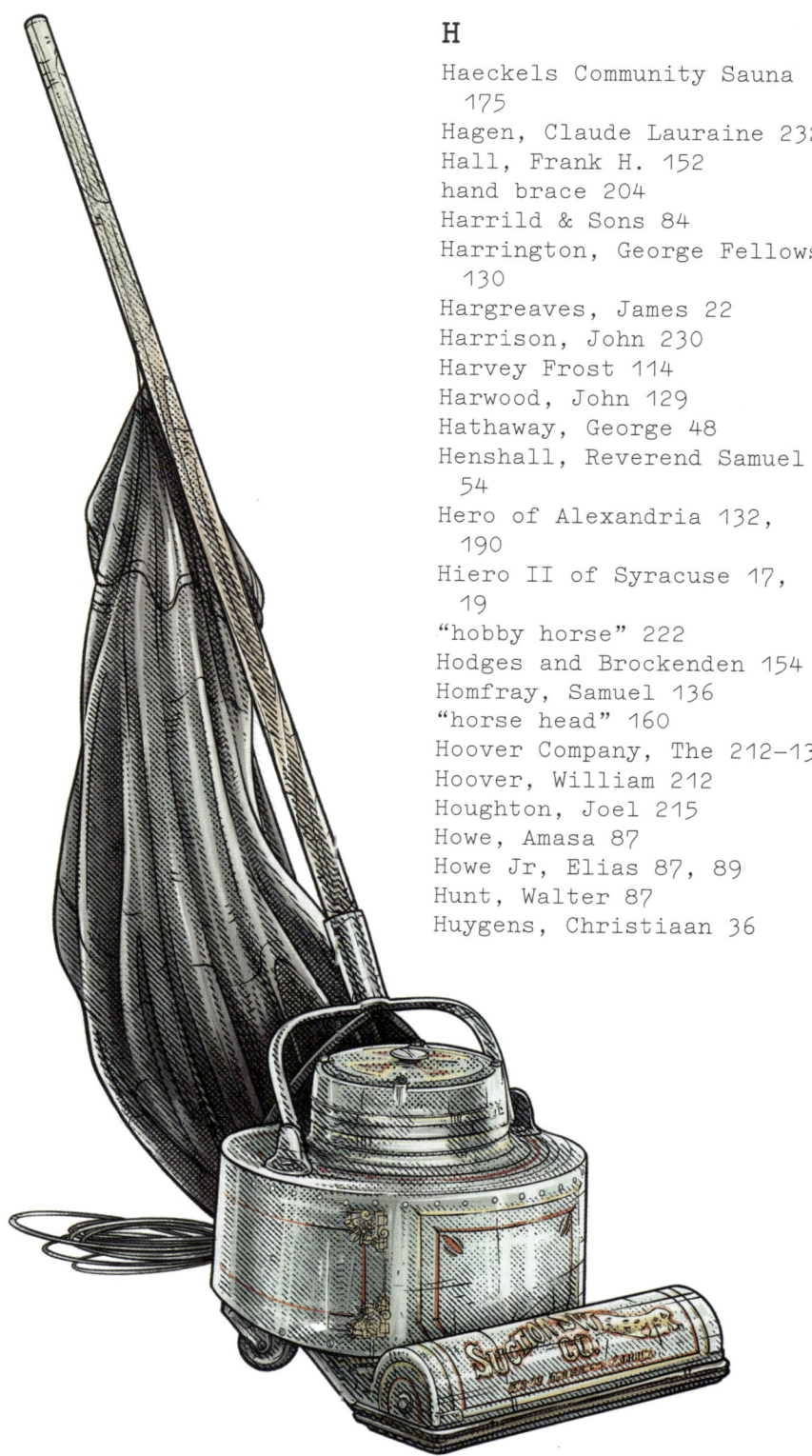

H

Haeckels Community Sauna 175
Hagen, Claude Lauraine 232
Hall, Frank H. 152
hand brace 204
Harrild & Sons 84
Harrington, George Fellows 130
Hargreaves, James 22
Harrison, John 230
Harvey Frost 114
Harwood, John 129
Hathaway, George 48
Henshall, Reverend Samuel 54
Hero of Alexandria 132, 190
Hiero II of Syracuse 17, 19
"hobby horse" 222
Hodges and Brockenden 154
Homfray, Samuel 136
"horse head" 160
Hoover Company, The 212–13
Hoover, William 212
Houghton, Joel 215
Howe, Amasa 87
Howe Jr, Elias 87, 89
Hunt, Walter 87
Huygens, Christiaan 36

I

Ideal School Supply Company 84
I. M. Singer & Company 89
Improved Strength Testing Machine Company 177
inclined plane 14
Industrial Revolution 22, 37, 48, 60–61, 70
Ingento No. 3 84, **85**
International Business Machines Corporation (IBM) 83, 158

J

J. R. & A. Ransome 218
Jackson, Charles Thomas 126
James Heeley & Sons 56
Jefferson, Thomas 62
John Charles & Company 104, 108

K

Kenney, David T. 212
Kenyon, Richard Walter 206
Kettering, Charles 146
Kochs, Theodore 142
Koken, Ernest 142
Koken's Congress Pedastal 142

L

Lallement, Pierre 222, 224
Lane, Allen 191
Lasceaux, Josef 171
Lassimone, Bernard 155
Lavadora 215
lever 15
Liverpool & Manchester Railway (L&MR) 138
Livesay, Charles H. 104, 108
Lochman, C. L. 116
Locomotion 138
Lofting, John 64
London Steam Carriage 136

Love, John Lee 156
Lowndes, Francis 228
Lufkin Foundry and Machine Company 160

M
Mabs, Harry 180
Macmillan, Kirkpatrick 224
Malouin, Paul-Jacques 62
Marcato, Otello 62
Massey Ferguson 27
Massiquot, Guillaume 84
Maudslay, Henry 79, 92
McAlpine, Robert 89
Messrs Reynold and King 182
Metcalf, John see "Blind Jack"
Meyer, Eugène 224
Michaux, Pierre and Ernest 224
Mill, Henry 80
Miller Falls 204
Miller, John 196
Mills Novelty Company 178
Model O 212-13
Monroe, J. F. and E. P. 57
Morrison, James B. 131
Morrison, William J. 170
Morse code 126
Morse, Samuel 126
Moss, Geoff 104, 107

N
National Cash Register Company (NCR) 145, 146
Nashtifan windmills 23
Nautilus 42-43
Newcomen, Thomas 132
No. 599 model 146

O
Olcott Climax Pencil Sharpener 156
Olivetti 83
onager 34

P
Palmieri, Luigi 125
Parkes, J. 178
Pascal, Blaise 38, 128
Pascal's Law 92
Pascaline 38
Pasteur, Louis 72
Patek Philippe 128
Patterson, James H. 145
Penguincubator 191
"perfect binding" 118
Perkins Brailler 152, **153**
Perpigna, Antoine 67
Peugeot 51-52, 65
pill silverer 110
Pinkert, Georg 192-93
Platform Gallopers 182
Polyphon Musikwerke 186
Pope, Albert Augustus 226
portable hydraulic riveter 99
post mill 23
Potter, Orlando B. 89
Powers, John Emory 90
Pratt, Joseph 80
Priestly, Joseph 66
Promenades Aériennes 196
Puffing Devil 136
pulley 16

Q
Quadricycle 163
Quakenbush, Henry 69
QWERTY keyboard 82

R
"R model" coffee mill 52
Rabkin, William 182
Rainhill Trails 138
Rampazetto, Francesco 80
Ranalah 99, 104-05, **106**, 107
Redgrave, Montague 180
Regina Company 186
Remington and Sons, E. 82
Reno, Jesse Wilford 150
"Revolving Stairs" 150
Riessner, Ernst Paul 186
Riggs, John 131
Ritty, James Jacob 145
"Ritty's Incorruptible Cashier" 145
rivet 94
RMS *Titanic* 94
Robert Stephenson and Company 138
Robert Tidman & Sons 182
"Roberval balance" 70
Roberval, Gilles 70
Rochester Time Recorder 158
Rocket 136, 138-39, **138-39**
rotary cork press 116
Rotatrim 85
Rotherham swing plough 27
Rover Cycle Company Limited 226
"Russian Mountains" 196

S
S. Maw, Son, and Thompson 110
SS *Archimedes* 140
SS *Great Britain* 140
"safety wheels" 196
Saint, Thomas 86
Salter, Richard 70
Savage, Frederick 182
Savery, Thomas 132
Saxon Wheel 22
Schweppe, Johann Jacob 66
Scott, Myron 196
screw 15
Seeberger, Charles 150-51
Sellers standard 15
Sellers, William 15
seltzer bottle 67
Setterington, John 174
"Sewing Machine Combination" 89
Sewing Machine War 89
shaduf 26
Shields, Dr Alexander 154
Sholes, Christopher Latham 80, 82

"Sign of the Orrery" 41
Singer, Isaac 87, 89
Singer Manufacturing Company 89–90
Smith, Billy 160
Smith, David A. 102
Smith, Francis Pettit 140
Smyth, David McConnell 118
Smyth Manufacturing Company 118
"Smyth Sewing" 118
Souder, Leamon 150
Soulé, Samuel W. 80, 82
Southend Pier 178
Spangler, James "Murray" 212
Speicher, Samuel S. 164
Spencer, Heath & George 228
spinning jenny 22
Spong & Co 58
spring scales 70, 71
Starley, James 224
St Louis World's Fair 170–71, 199
"Stairway" 150
Stephenson, George 136, 137
Stephenson, Robert 137, 138, 139
Stevens, John 140
Stylixynon 155
"sucking worm engine" 64
Surge Milker 154
swan neck 64
Switchback Railway 196
"Symphonion" 186

T

taille crayon 155
Tainter, Charles Sumner 219
Tasker, Robert 227
"telharmonium" 83
Thimonnier, Monsieur 86–87
"Thistle", the 154
Thurman, John S. 210
Thomas Linley, Sons & Co of Sheffield 35
Thomson, LaMarcus 196
Tosty Rosty 173
Treatise on the Steam Engine, A 98
Trevithick, Richard 136
Troughton, Edward 41
Troughton, John 41
Trout, Walter 160
Turtle 42, 140
Turtle-back 90
Tweddell, Ralph Hart 99
"Type M" Ranalah 108
typewriter
  Daugherty "Visible" 83
  Electromatic 83
  Remington No. 1 58, 82
  Remington No. 2 58, 82
  Sholes and Glidden 82
  Underwood Model One **81**, 83
  Underwood Model 5 83
type-mould 28

U

Underwood, John T. 83
USS *Princeton* 140

V

Vail, Alfred 126
van Bercken, Lodewyk 109
Van Stoeser, W. 180
Victoria, Queen 50, 58, 84, 127, 174–75, 177, 206
Vitantonio, Angelo 62
Vivian, Andrew 136
Volta Laboratory 219

W

Wagner, Franz X. 82–83
Wall Street Crash 129, 176, 215
Warner, Ezra J. 73
Warwolf 32
water screw see Archimedes' screw
"water velocipede", 192
Watt, James 132, 136
Weald & Downland Living Museum 20, 35, 133
wedge 14
Wells, Horace 131
Wharton, John C. 170
Wheatstone, Charles 126
wheel and axle 26
Wheeler, George 151
Whitworth, Sir Joseph 15
Wiesenthal, Charles Frederick 86
Wilcox & Gibbs sewing machine **88**, 90–91
Wilcox, James 90
Wilhelm I, Kaiser 128–29
Williams, Harry 180
Wizard Fortune Teller, The 179
World's Colombian Exposition 173, 198

Y

Yun-ŭi, Ch'oe 28

Z

Zoltar 178

ABOUT THE AUTHOR

Dominic Chinea is a craftsman and expert on the BBC's much-loved show, *The Repair Shop*. Originally starting his career in graphic design, Dom's passion for vehicles and metalwork led him to jobs in classic car restoration and set design, before joining *The Repair Shop* team. Dom presents the BBC show, *Make It at Market*, helping entrepreneurs to harness their crafting skills and start a career.

You can find Dom's latest restoration projects on his YouTube channel. He is the author of *Tools*, revealing the history and uses of the most inspiring tools from his workshop.

ACKNOWLEDGEMENTS

Making a book like this is a real team effort, and throughout the process I've been lucky enough to get support from some incredibly knowledgeable and talented people. I want to thank all of you who've helped me with my research and taken time to speak about some of the lovely old machines featured in the book.

PICTURE CREDITS

All illustrations by Lee John Phillips.

Photography on pages 2, 8, 96, 134, 208 and back cover by Jack McGuire.

The publisher would like to thank the following for their kind permission to reproduce their photographs:

(Key: a-above; b-below/bottom; c-centre; f-far; l-left; r-right; t-top)

51 Image Courtesy National Gallery Of Art, Washington: Index of American Design (br). 52 Image Courtesy National Gallery Of Art, Washington: Index of American Design (b). 55 Alamy Stock Photo: Chronicle. 67 Alamy Stock Photo: Alistair Scott (b). 68 Getty Images: Universal Images Group / Universal History Archive (b). 185 Alamy Stock Photo: The History Collection (br). 186 Getty Images: Archive Photos / Jay Paull (b). 188 Getty Images: Archive Photos / Jay Paull. 207 Alamy Stock Photo: Marcus Harrison - adverts (tr). 211 Alamy Stock Photo: f8 archive. 225 Getty Images: Science & Society Picture Library. 228 Getty Images: Universal Images Group / Universal History Archive (br)

All other images © Dorling Kindersley Limited